Grapes Cultivation

Grapes Cultivation

Dr. Savita Kumari
Subject-Matter-Specialist (Home Science)
Krishi Vigyan Kendra
Lahaul & Spiti II, Tabo–172 113 (Himachal Pradesh)

SSPH

SATISH SERIAL PUBLISHING HOUSE
403, Express Tower, Commercial Complex
Azadpur, Delhi-110033 (India)
Phone : 011-27672852, Fax : 91-11-27672046
e-mail : info@satishserial.com, hkjain1975@yahoo.com
Website : www.satishserial.com

Published by :

SATISH SERIAL PUBLISHING HOUSE
403, Express Tower, Commercial Complex, Azadpur, Delhi-110033 (INDIA)
Phone : 011-27672852 Fax : 91-11-27672046
E-mail : info@satishserial.com, hkjain1975@yahoo.com

ISBN 978-93-88020-11-4

Composed, Designed & Printed in India

List of Tables

List of Figures

List of Plates

Contents

Chapter 1

Introduction

Grape (*Vitis vinifera L.*) is an important fruit crop in India. Grapes are the third most widely cultivated fruit after citrus and banana. Major grape-growing states are Maharashtra, Karnataka, Andhra Pradesh, Tamil Nadu and the north- western region covering Punjab, Haryana, Delhi, Western Uttar Pradesh, Rajasthan and Madhya Pradesh (Singh, 2010). In Haryana, grapes are cultivated in an area of 111.00 (000 ha) with a total production of 1235.00 (tons) and productivity of 11.10 (tons/ha) in 2010-11 (National Horticultural Board, Government of India). Haryana is the sixth largest producers of grapes in the country with 5.7 million ton/hectare/year. Grapes cultivation is grown under a variety of soil and climatic conditions (Shikhamany, 2001). In Haryana, it is usually cultivated from January to June end. Near about 75 per cent of grape produce is used for table purpose, 15-20 per cent is dried for raisin production and 5 per cent is used for manufacturing of juice and wine. Consumption of grapes is beneficial to health, in addition to the usual nutrients grapes are also rich in phenolic compounds and other natural antioxidants. The export percentage of grapes is not more than 4 per cent of total production. According to the International Labour Organization, the agricultural sector is one of the most hazardous sectors. Agriculture work were possesses several tasks they are risky for health exposure to the weather, close contact with animals and plants, extensive use of chemicals and biological products, difficult working postures and lengthy hours, and use of hazardous agricultural tools and machinery. Agricultural workers involve several strenuous activities like ploughing, spading, carrying, uprooting, planting, weeding, cutting, shafting, threshing, sweeping, etc. Musculo-skeletal disorders were common among farmers. Farmers handle heavy workloads often in awkward posture and experiencing some work related problems. They experience high rates of low back, shoulder, hand, knee and upper extremity disorders (Donald, 2006). Grapes cultivation is one of the agricultural activities. Grapes cultivation involves various activities like land preparation, irrigation, manuring, pruning, harvesting, transportation, etc. Grape production is very labour intensive operation i.e. Grape vineyard workers faces high stress on the hands during pruning of the grapevines under highly repetitive conditions (8 to 10 week period of intense and fast-paced work) and also the cumulated duration of exposure over the entire day was high, i.e. approximately 8 to 10 hours per day over a 4-month period. The frequencies of cuts, the pruning task were highly repetitive, and vineyard worker more autonomy than industrial workers and can modulate their rhythm of works and takes break they need. The use of pesticides and fertilizers in grapes cultivation is a common everyday practice. It was found that pesticide crop sprayers were more likely to report work-related skin symptoms, asthma and other respiratory health problems among farm workers (Jeebhay *et al.* 2002). Labour-intensive crops, such as fruits and vegetables, are treated extensively with pesticides. Most (more than 50%) of farm workers are hired for harvesting operations, they might be exposed to pesticides thereby, several skin diseases, poisoning-abdominal pain, nausea, dizziness, vomiting, headache, eye problems, such as rashes, inflammation, or corneal ulceration had been seen. Pruning and harvesting was one of the most expensive and labour consuming process. Several manual tasks performed with a cutting tool, such as tie vine shoots, spring pruning, pruning, remove grapes, harvesting. There was stereotype work movements of hands are made upto 37 thousand per a shift. Pruning imposes high biomechanical strains on the hand–wrist and repetitiveness of the task. Physical exposure during pruning activity the high prevalence of hand disorders among vineyard workers. The use of ergonomic pruning shears is advised to lower force exertion and to reduce the frequency of awkward wrist postures during pruning. Vineyard workers may risk back and the upper extremity injury during occupational tasks, such as harvesting, pruning, and shovel weeding due to the postural demands and repetitive performance associated with these activities.

Many tasks, such as dormant pruning, shoot suckering and crop harvesting were done repetitively by hand and could result in musculoskeletal disorders (MSD) among the workers. Pruning had also been associated with increased risk of developing cumulative trauma disorder of the wrist among workers. Vineyard rows (about 30 feet long each) was planted 8 to 12 feet apart, with about five vines per row. Pruning one vine takes about 60 seconds. Pruning work were shifts are 8 hours, with approximately 2400 cuts per hour i.e., about 60 vines per hour, or 480 vines per day (Roquelaure *et al.* 2002). In vine pruning and harvesting, the operators are open to a great deal of strain likely to trigger or accelerate the Work related musculoskeletal disorder generating mechanisms including repetitive movements combined with external forces acting on the finger, hand wrist system and static work in the upper arm shoulder system. Pruning of wine grape vines have also been found to have awkward postures of the shoulders. Hand pruning of vines requires the frequent squeezing of the hand to engage a clipper, or the frequent use of a knife. Hazards from the knife are obvious, as there is no solid surface against which to place the vine, shoot or stalk and frequent cuts to the fingers, hands, arms, legs and feet are likely to result. Clipper is the preferred tool for pruning; the major safety hazard is the threat of cuts from contact with the open blade while placing a vine or stalk in the jaws, or from inadvertent cutting of a finger while also cutting a vine or stalk. The frequency and effort required for cutting determines the likelihood of development of cumulative-trauma injuries. The force required to operate a common clipper is in excess of recommended values, and the frequency of effort indicates the potential for cumulative-trauma disorders, according to accepted guidelines. Lifting and carrying of loads is typically associated with harvesting of grapes, which are usually hand harvested and carried in some type of basket or carrier to the edge of the field, where they are collected. Loads are often not heavy (10 kg or less), but the distance to be travelled is significant in many cases and over uneven terrain, which may also be wet or slippery. Training plants on trellises requires the exertion of substantial force, a force that is increased by the weight of the vines, foliage and fruit. This force is commonly exerted through the arms, shoulders and back, all of which are susceptible to both acute and long-term injury from such overexertion. Grapes are also commonly cut from the vine during hand harvest. This cutting motion is also very frequent (hundreds of times per hour) and requires sufficient force to cause concern regarding cumulative-trauma injuries. Harvesting is often done in awkward postures and at a rapid pace. Persons typically twist and bend, bend without bending the knees and move quickly between the bush or vine and the container. Containers are sometimes placed upon the ground and pushed or pulled along with the worker.

Several studies have already been done and more are on the way to develop a perfect ergo solution for pruning and harvesting of grapes. A hook was used as a supporting device for carrying fruit, is an easy solution. Various ergo-solutions have already been developed for the activities like pruning, weeding and harvesting. However, when full, the bucket had to be lifted almost at shoulder level, which is stressful. And also a small ladder allows a good working posture during harvesting. A light-weight ladder was preferred the ladder had to be moved from one place to another (lifted) so frequently. So, to avoid excessive reaching and acrobatic postures: Moving the harvested tree fruit in a picking train is an efficient practice. The fruit was placed directly from the tree in a pallet box. Work is often combined with a picking platform that is used for the higher localized fruit. For grape harvesting, alternative knife-handles were used showed worker acceptance of an improved grip for the combination pinch and power grip. The pinch grip is necessary for quick and precise knife placement, and the power grip is necessary for strength needed to jerk the knife through the stem of a grape cluster. Though the existing knives could not be identified as causing any reported injuries other than minor cuts to the non-dominant hand workers was continuselly using of the alternative handle. Nursery a worker was disability due to carpal tunnel syndrome from excessive use of cutting shears was able to return to their propagation cutting task due to the powered shears. In weeding, motorized tools were minimizing the problem for trunk bending and twisting. Growing plants at waist level allows the worker to adopt a more upright posture. The posture for pruning is relatively easy. The tools used were simple secateurs, scissors or a knife with a sharp blade. Long handle enable workers to reach greater distances while keeping a normal posture. Two hands were used to increase the power and shares the strain on both arms and hands. Using pruning clipper or lopping shears with long handle. Long handle avoids the need to kneel or squat at ground level. Pruning saw with a short handle work the fastest and feel the most stable and accurate. An angled handle allows a higher force exertion. Apple pruning using secateurs were one of the handles has an outer surface rotates as the worker's grip draws the two arms together, then decreases the strain on the fingers and hand.

All the above mentioned developed ergo solutions were not available in Haryana and moreover, they are also expensive hence, locally available need based ergo-solutions is the need of an hour.

Therefore, in light of all these considerations/factors, the research was carried out focusing on the following objectives:

1. To study work profile and working conditions in grape cultivation.
2. To assess the occupational risks in grapes cultivation.
3. To develop an ergo–solution for highly risk prone activity in grapes cultivation.

❑❑❑

Chapter 2

Review of Literature

A comprehensive review of literature is essential in any research endeavour. The main function of the review of literature is to make present research up-to-date with the existing research work in the field investigation. The brief resume of past literature and researches, relevant to the present study, has been incorporated in this chapter. The pertinent literature has been presented under the following sub-heads:

2.1 Work profile and working conditions in grapes cultivation.

2.2 The occupational risks in grapes cultivation.

2.3 Ergo-solution for highly risk prone activity in grapes cultivation.

2.1 Work Profile and Working Conditions in Grapes Cultivation

Bonger *et al.* (2002) reported an association between at least one work related psychosocial factor and adverse upper extremity symptoms or signs. High perceived job stress was consistently associated with all upper extremity problems (UEP) in high and lower quality studies. Although not often studied, non-work related stress was also consistently associated with UEP. In addition, high job demands were also in most studies associated with these disorders.

Saha (2003) suggested two types of rest pause, one for light work and the other for heavy work viz., 10 min of rest followed by 30 min of light work and 30 min of rest followed by 10 min of heavy work. It was shown by a mathematical computation that such arrangements of work and rest periods would not only keep the level of physiological strain at an acceptable limit, but also high production with less physiological cost.

Monarca *et al.* (2003) analyzed the human heat balance in the different conditions (from 20 to 45 C^{o}) and found that though the thermal power given off by work activity remained constant, the power given off by body, in different ways, varied according to the air temperature variations. In welfare conditions, the maximum loss of energy occured by perspiration (37%), by convection (15%) and by diffusion from skin (11.5%). The energy loss in other way was about 10%. But the higher limit of welfare parameters (25°C), the power given off by perspiration rose till 43.6%. Perspiration had a very important role to guarantee the human heat balance in stress conditions. Water given off in the form of perspiration or breathing rose with the increase in temperature. The amount of water given off by perspiration increased and was above 300 g/h, when the air temperature was over 40°C. Undoubted, the water loss produced an alteration of the body water equilibrium with a consequent loss of minerals, leading workers to severe health risks.

Monaria *et al.* (2003) reported that heat stress was common problem in Walk-In-Tunnel (WIT) polyhouse due to extreme temperature. Water given off by perspiration increased and was above 300g/h, when the air temperature was over 40^{0}C. Undoubted, the water loss produced an alternation of the body water equilibrium with a consequent loss of minerals, leading workers to severe health risks and could generate some problems of breathing.

Van *et al.* (2005) studied to determine the individual and work related factors which were associated with performing modified work and to evaluate the influence of modified work on the duration of sick leave and health related outcomes among the employees with musculoskeletal complaints. The study was conducted for 12 months, in this perspective study, total of 164 employees who remained on sick leave for 2-6 weeks due to musculoskeletal complaints were less likely to perform modified work, when their regular work was characterized by frequent lifting and their relationship with colleagues was less than good. Employees were more likely to return to modified work, when they had a better mental health, had prolonged periods of standing in the regular job and had less skill complaints and disability but not modified by work. They considered that modified work did not influence the total duration of sick leave for employees or sick leave due to musculoskeletal complaints.

Hagberg and Sundelin (2006) proposed benefits of rest intervals and reported that frequent rest intervals could assist in reducing the perception of postural discomfort and musculoskeletal disorder.

Further postural variation and a break away from constrained posture could be effective in reducing or decaying the experience of musculoskeletal discomfort.

Kjellstrom (2009) reported that carrying out prolonged physical activity in a hot, humid environment increased the risks of heat exhaustion and heat stroke. In addition to these symptoms, excessive heat exposure could cause heart and kidney problems. Exposure to a hot environment also changed physiological parameters, such as pulmonary ventilation, vasodilation, sweating and blood flow, thereby causing increased absorption of xenobiotics through pulmonary or cutaneous routes.

Gurav and Salunkhe (2011) studied problems and prospects of women in grape farms and found that 63% respondents did not get any incentives and 79% respondents were not satisfied with their wages. At the same time, 52% respondents mentioned that they got 8 months job guarantee in the vicinity of their residence, 54% respondents got refreshment facility and 67% respondents got transport facility to travel from their residence to grapes farm i.e. working place. 67% respondents were not aware about cleanness and hygiene, 84% women employees worked as a labour and farmer, 53% got medical check up facility, 14% were psychologically harassed, which was not good on other part, and 92% women employees got training facility.

Poupart *et al.* (2013) found that the potential effects of heat exposure on occupational health and safety were both direct and indirect. Exposure to high ambient temperature caused an increase in body temperature, which translated into cutaneous vascular dilation, sweating and increased heart rate.

Dahiya *et al.* (2013) found that in walk-intunnel (WIT) and natural ventilated polyhouses (NVPH), the temperature was 69.5 and 52.29% higher and the humidity was 96.37 and 85.19% higher in comparison to open farming in the months of January and May. No significant difference was found in temperature, humidity, dust, solar radiation and CO_2 level between open and anti-insect net shade house (AINH). In Hi-tech polyhouse, the environment was totally controlled by computer and was not found to much strenuous. The CO_2 level in WIT was found to be 359.82 and 332.95% higher in comparison to open farming. According to mental workload scale the working condition in WIT and NVPH was found 'risky' the score of 51.17and 47.6%.

Vinayak *et al.* (2015) conducted a study on attitude of farmers towards grape cultivation and export. They earned average Rs13.3 lakh per annum by exporting their produce. Average land holding was 10 acres and productivity of grapes obtained was 11.7 ton /acre. Farmers had more favourable attitude towards grapes cultivation as a result of their association with Mahagrapes. Total annual income, risk taking behaviour and achievement motivation of members were positively and significantly related to attitude towards grapes cultivation and export. Thus, the study implied that farmer's organization like Mahagrapes helped in developing positive attitude among farmers towards grape cultivation and export, motivating them to earn more income annually by exporting their produce.

2.2 The Occupational Risks in Grape Cultivation

Meyers *et al.* (1999) conducted a study and found that awkward postures in the harvest task included moderate shoulder flexion and mild shoulder abduction frequently severe forearm pronation and occasionally severe forearm supination moderate-to-severe ulnar deviation and virtually no radial deviation of the wrist. Trunk flexion ranged from mild during cutting to severe during stooping to gather fallen grapes. Neck flexion was occasionally severe but often the neck was in extension of more than 5^0. Workers were often observed to be in a forward head position of trunk flexion with simultaneous neck extension. Additional risk factors included constant ulnar deviation of dominant hand, very high energy cost, pressure on hand from poorly-designed knife handle cuts on non-dominant hand from knife and long carried with bins held overhead for some work methods.

Meyers *et al.* (2000) studied that stress on the lower extremities occurred, when carrying heavy loads (e.g. bins, bags, or baskets of vegetables or fruits) from the field to the truck. High knee stress occurred, when the knee was flexed during long static postures, such as squatting to pick weeds or crops. Kneeling can also create contact stresses that may lead to knee discomfort and disorders. Unidentified risk factor may be prolonged standing on hard, uneven, or unstable floors (e.g. muddy fields, concrete slabs, etc.).

Meyers *et al.* (2000) reported that nursery workers were exposed to awkward postures and poor lifting conditions in several of their activities, handling of plants in 1-gallon containers, pruning of plants, weeding, plant labelling and loading/unloading trucks.

Wakula *et al.* (2000) conducted a study to find out the stress- strain in the field and laboratory workers. Study showed that pruning grapevines and wine harvesting involved a combination of dynamic and sensorimotric work, and also a high incidence of ergonomically undesirable posture of the trunk, the upper limbs and the head irrespective of cutting tools used.

Mines *et al.* (2001) reported that individuals working full days under stressful conditions were more prone to accidents. When injuries occur, they can be severe. Workers perceived that working with badly maintained tools and machines posed considerable risks to their health. Poor worksite maintenance was particular concern to greenhouse workers whose jobs involved standing on electric carts with adjustable platforms some six to eight feet off the ground.

Roquelaure *et al.* (2002) studied that the biomechanical strains on the hand wrist system were high during grapevine pruning. The high level of strain in the hand disorders was observed in winegrowers. The study showed the most worker scope with the biomechanical load by adopting working strategies that reduced the occurrence of extreme wrist positions combined with high force exertion. Nevertheless, the introduction of electric pruning sheared or ergonomic hand-powered pruning sheared, in association with adaptation of work organization, needed, to lower force exertion and reduce the frequency of awkward postures of the wrist during pruning.

Murali *et al.* (2004) conducted an investigation to assess the angle of postural deviation of body of farm women while performing the selected farm activities using traditional methods and improved tools and their relation with heart rate and perceived exertion. Findings revealed that there was slight variation in the angle of body bend at cervical and lumbar region of women while performing the selected activities using traditional methods and improved tools. However, the correlation between angle of body bend, heart rate and perceived exertion was non-significant for most of the activities.

Chapman *et al.* (2004) reported the evaluation of an intervention to reduce musculoskeletal hazards among fresh market vegetable growers. The intervention was disseminated information about the practices to growers through trade publications, public events, university extension, and growers already using the practices. A mail questionnaire was administered to vegetable growers before and after the intervention. Strawberry growers were used as a comparison group and also received questionnaires. The results revealed that after the intervention, more vegetable growers reported seeing information about mesh bags in trade publications (37% vs. 59%) and information about standard containers at public events (33% vs. 49%). Levels of self-reported adoption increased for containers (38% vs. 54%) and approached significance for bags (8% vs. 17%). Grower perceptions of bag profitability improved (2.6 vs. 3.8). Strawberry grower control results were unchanged. Better information flow to growers may be able to increase the speed with which agricultural practices with better ergonomics were adopted, especially when the practices are more profitable.

Chatzi *et al.* (2005) demonstrated the high prevalence of allergic rhinitis and work-related respiratory symptoms in grape farmers compared to control subjects. It was also suggested that grape farming was possibly associated with increased allergic sensitization to specific pollens, low baseline FEV1 (Forced Expiratory Volume) and increased bronchial hyper-responsiveness.

Fathallah *et al.* (2006) conducted a study to evaluate the effectiveness of an intervention on the incidence of musculoskeletal symptoms among workers during two grape harvest seasons. There was five-fold reduction in workers post seasons musculoskeletal symptoms scores, without significant reductions in productivity.

Youkmi (2006) reported that viticulture workers were at risk of work-related musculoskeletal problems, especially of the wrists and hands, from vine pruning work, and could develop allergic diseases, including occupational asthama, from exposure to insect -pests growing on vines. They may also be at risk of illness from exposure to the many classes of pesticides used to keep vine pests in check. Wine production workers also faced health risks. The most serious of these working in confined spaces with low oxygen and high carbon dioxide levels could cause death. Work-related injuries were more commonly reported than work-related diseases in the wine industry, physicians assessing the medical problems of wine industry workers should be aware of the occupational health risks faced by these individuals.

Kato *et al.* (2006) studied relative risk of developing MSDs among five of the common trellis systems used in the wine grape industry. The study showed that the lyre and VSP 4×4, both of which encompassed relatively extreme trunk postures, resulted in the largest wrist flexion angles compared with the other trellis systems. The relatively large trunk flexion and extension angles in combination with increased wrist flexion/extension observed in these systems increased the risk of developing MSDs of both the wrist and the back. It was observed that the VSP system resulted in both the most time spent in a neutral trunk posture combined with acceptable wrist postures.

Sriwarno *et al.* (2006) found that repetitive movement (10-12), high pinch force (in tying the thread to wire and plant), frequent trunk flexion and extension motion, cutting finger (due to thread) and high eye strain were major problems in workers of tying activity in polyhouse farming. Workers of tying activity were found to be having severe problems in neck, shoulder and low back. Posture of workers was found to be the most dangerous due to excess repetitive movement and extension of shoulder and head.

Jeebhay *et al.* (2007) carried out a cross-sectional study on 207 workers employed in grape farms in the Western Cape. The prevalence of work-related wheezing (26%) was much higher than other allergic symptoms and 9% of workers reported physician-diagnosed asthma. When compared with general symptoms, the prevalence of work-related symptoms was almost comparable for wheezing (26/30), rhino conjunctivitis (24/25) and urticaria/skin symptoms (15/20).

Hisataka *et al.* (2007) conducted a study on orchard farmers' musculoskeletal symptoms while bagging pears and apples. The prevalence of stiffness and pain in the neck and shoulder, muscle tenderness in the shoulder regions, and pain in neck motion were found to be significantly higher, when bagging pears than apples. Musculoskeletal symptoms of parts other than the neck and shoulder did not differ between the two types of bagging. The working posture of elevating the arm more than 90° was assumed to account for 75% of the time bagging pears, against 40% for bagging apples. Overhead work requiring arm elevation and head extension was considered to be closely related with shoulder-neck disorders among farmers.

Pezzi *et al.* (2009) conducted an experiment on the cutting force of vine canes during vegetation rest was a parameter which depends on vine variety, branch thickness and pruning period. High values of cutting force can generate tiredness of operator in manual pruning or cutting damages during mechanical pruning. A percentage of damages higher than 2-3% was certainly to be avoided for the sanitary problems that they can cause to the permanent cordon. Also an incidence of 20-25% of light damages was to be considered negative, not only in the infrequent case of totally mechanical pruning, but also because it slows down the hand finishing which usually follows the mechanical operation. The results can lead to an improvement of the optimization of the pruning times scheduling and to an increase of mechanical efficiency according to the vine type. When the mechanical pruning was necessarily to be made during the vegetation rest, when the wood was harder, it was better using a higher oscillation frequency of the blades.

Kumudini and Hasegawa (2009) reported the workload and awkward postures among strawberry farmers. Questionnaires, direct observations and intertwining methods were used to collect the data. Low-level strawberry beds and long working hours were recognized as critical factors that led to farmers' health problems including suffering from low back pain, heavy fatigue, unsteady feeling after work, etc. There was a substantial increase in fatigue in busy seasons compared with slack seasons. When compared with the other workers, such as office, technical, sales and blue-collar workers, female farmers' general fatigue level was considerably high. Findings indicated that strawberry farming was a stressful occupation and farmers were strained under heavy work load, monotonous and repetitive work and poor working conditions with frequent fatigue symptoms and severe influence on health.

Casado *et al.* (2009) performed a complete tasks analysis during the annual cycle in 11 vineyard farms, located in different regions with a total of 162 farm workers. The intrinsic risk index of upper limb injury was calculated for every task, specifying the temporal patterns of annual exposure for each professional profile. The Occupational replication Action OCRA checklist score obtained was 19.9. The prevalence of 19.8% exposed workers was affected by one or more upper limb WMSDs.

Montomoli *et al.* (2009) reported the ergonomic analysis using OCRA check-lists and NIOSH method to study the spine and upper limb work-related musculoskeletal disorders in grape orchards. Each task analyzed showed a high risk of biomechanical overload of the upper extremities. Regarding the manual material handling in the grape-harvest, band of risk changed from yellow to green. Considering the total number of the exposed workers studied in risk assessment only 32% of them were allowed to submit to a clinical examination by a specialist in rheumatology. The most relevant pathologies were the wrist and fingers tendinitis and the carpal tunnel syndrome.

Brumitt *et.al* (2011) showed the musculoskeletal health of Latino migratory vineyard workers. Nearly half (48.4%) of all vineyard workers reported experiencing musculoskeletal symptoms (MSS) in at least one region of the body. The primary region of reported MSS was the back (32% of all men and 43.7% of all women). Many workers reported that their pain was associated with the physical demands of their employment.

Bernard *et al.* (2011) studied biomechanical and psycho social work exposures and musculoskeletal symptoms among vineyard workers. This cross-sectional study was based on a random sample of 2,824 males and 1,123 females vineyard workers in France. Neck/ shoulder, back and upper and lower extremity symptoms were evaluated using the Nordic questionnaire. Biomechanical exposures included 15 tasks related to vineyard activities. Psycho-social work factors included effort-reward imbalance and over commitment, measured using the effort-reward imbalance model, and low job control and insufficient material means. Further results found that pruning-related factors increased the risk of upper extremity pain for both the genders, of back pain for men and of neck/shoulder and lower extremity pain for women. Psycho-social work factors, which were insufficient material means, over commitment (both the genders), effort-reward imbalance (men) and low job control (women)were associated with musculoskeletal symptoms, back and upper extremity pain for both the genders and neck/shoulder and lower extremity pain for men.

Callea *et al.* (2014) conducted a study on occupational illnesses related to physical strains in apple harvesting. The results found that neck and dorsal pains and fatigue caused for each operator. The VO_2 was equal to 82.33±27.40 lO2/h for women and 67.00±27.60 Lo_2/h for men. The heart rates were of 115±6.00 bpm for women and 113±5.65bpm for men. The VCO_2 was of 63.81±21.45 lCO_2/h for women and 45.10±25.53 lCO_2/h for men, while energetic equivalent and body surface area was similar for both the genders, about 5.60W×h/Io_2 and 1,80 m2 on average. Women's metabolic rate had a very high value over 290W×m^2, although for the men it was between 200-260W×m^2. According to OWAS, low apple picking was ranked in class 2, high apple picking in class 1, and apple transportation belonged to class 3.

Ng *et al.* (2015) investigated the associations of musculoskeletal disorders (MSDs) with socio-economic background, occupational exposure, social lifestyle and postures adopted during harvesting tasks. OWAS was used to determine the severity of awkward posture based on videos of harvesting tasks recorded for each respondent. Results found that increasingly educated respondents had higher risk of developing MSDs. Shorter daily work duration and longer resting duration appeared to increase the risk of neck and shoulder disorders among harvesters, which may be attributable to organizational work design. Awkward posture was a particularly significant risk factor of MSDs among FFB(Fresh fruit bunches)collectors occupational exposure, postures and certain socio-demographic backgrounds explained some, but not all, the risk factor of MSDs among harvesters.

2.3 Ergo-solution for Highly Risk Prone Activity in Grape Cultivation

Meyers. *et.al* (2000) found that a movable table should be used to elevate trays while weeding. Workers stood while weeding, relieving prolonged stress on back. Workers must still stoop to pick up and replace trays, but change of position was better than prolonged position. Tray was closer to worker, reducing strain from reaching. Strain reduction would improve worker performance. He found that a smaller, lighter tub that had add-on grips and weighed an average 46 pounds, when full. The lower weight was easier on the back, knees and arms. The narrower width positioned the tub's center of gravity closer to the worker, which reduced stress on the back. Lighter weight and smoother bottom surface reduce the sideways forces on the knees when pushing the tub down the row. Better handles reduce pressure points on the fingers.

Richardson *et al.* (2005) studied the ergonomics intervention to reduce back strain among apple harvest workers in New York State. The use of hip belt did not significantly alter time spent in various postures: 78.6% of interviewed workers preferred the modified bag, 71.4% noted a difference in the back, neck, or shoulder, while 64.3% said regular use of modified bag would slow their work. Re-design of the hip belt-bag attachment mechanism was needed, so that it was easier to hook, and the bag could move around the worker, and so that there were no extra protrusions to catch on branches, fully developed orientation and training for bag users was needed, to assure that a clear understanding of the mechanism of weight transfer was understood, and so that bag adjustments were not time consuming.

Peppelman *et al.* (2006) reported that moving the harvested tree fruit in a picking train was an efficient practice. The fruit was placed directly from the tree in a pallet box. This type of work was often combined with a picking platform that was used for the higher localised fruit. He also reported that Frontloader the pallet was loaded or unloaded with the driver/operator maintaining a neutral position of the spine. Frontloader was that for good vision, when driving forward less pallet was carried.

Skelton (2007) found that to avoid excessive reaching up-turned light-weight crates were used for apple harvesting. The handle, a half broom stick was attached to reduce bending of the back, when moving the crates. A crate was also used to bring the bucket or other basket to a higher level. However, old orchards sledge was used as a table to put the bucket or crate on. Stooped postures were avoided.

He also reported that a hook was used as a supporting device for carrying fruit and was easy-to-use solution. However, when full, the bucket had to be lifted almost at shoulder level, which was stressful. He also reported that a small ladder allowed a good working posture during harvesting. A light-weight ladder preferred to be moved (lifted) frequently also using a ladder was observed to avoid excessive reaching and acrobatic postures. He also reported that picking platform was engine-powered or pulled forward by a tractor. The fruit was picked until shoulder height while standing on the ground. Afterwards, the higher fruits were picked while standing on the driving platform. Working above shoulder was diminished and people stood comfortably. The wooden crates were placed on the vehicle i.e. shorter distance for emptying the picking basket.

Kaur and Sharma (2010) conducted ergonomic evaluation of vegetable plucking activity with traditional (ordinary knife) and improved tool (ring cutter). Ergonomic assessment of both the methods showed that by using ring cutter physiological and muscular stress of workers in terms of heart rate, energy expenditure rate, and physiological cost of work and grip fatigue was reduced as compared to traditional method. Thus, new tool i.e. ring cutter was found to be beneficial to improve work efficiency of farm women.

Romano *et al.* (2012) conducted a study on anthropometric parameters and analysis of stress produced during manual pruning in vineyard. Experimental tests were conducted in the laboratory on branches from vineyards during winter pruning on a group of 10 healthy volunteers subjected to tests using cutting shears mainly different for characteristics of the handle and blade. The study recorded for each individual: age, weight, height, length, width and thickness of the hand volume, length, width, volume and thickness of the arm and forearm. During the cutting operation effort of the hand was detected in terms of intensity and duration by a shear equipped with sensors. A relationship was established between anthropometric parameters and stress as a function of the hand grip of the different types of cutting tools.

Parish *et al.* (2013) conducted an experiment on comfort analysis of labour during different tea-leaf harvesting practices. They found that during all operations, heart rate of a labour increased at faster rate in the beginning of the experiment, then stabilized for certain time of operation and finally reached to maximum value, when labour felt tired. Labours indicated their fatigue after 30, 25 and 35 min of operation in manual, shear assisted and motorized shear assisted harvesting methods, respectively. Motorized shear assisted harvesting method presented longer time of operation compared to manual and shear assisted harvesting methods. Moreover, average heart rate values and oxygen consumption rate were observed lowest for motorized shear cutter assisted harvesting, compared to those of shear assisted harvesting and hand plucking.

Lee *et al.* (2013) concluded that agricultural work involved occupational risk exposures. Fruit cultivation was one of the most hazardous crops. Grapevine cluster thinning task involved activities related to the main risk factors associated with upper limb work-related musculoskeletal disorders. An experiment was designed to test the working chair. A grapevine's working space was built for the experiment and working chair was installed on it. The muscle activity and range of motion subjects were used to the working chair for 30 min for each experimental condition. Decrease in muscle activity was found in all leg muscles and significant decrease of muscle activity was found in left Gastrocnemius. The range of motion of the trunk and knee also decreased, when working chair was used. Discomforts in lower back, thigh and shank region were reduced significantly. However, in upper limbs, muscle activity tended to increase in working chair compared with conventional task.

Thiyagarajan *et al.* (2013) conducted an experiment on ergonomic evaluation of four models of sugarcane harvesting knives. The maximum aerobic capacity of the selected 10 individuals varied from 1.84 to 2.19 l/min for sugarcane harvesting. The heart rate and oxygen consumption rate of the sugarcane knives varied from 132.55 to 138 beats/min and 1.171 to 1.253 L min-1, respectively. The energy cost of sugarcane harvesting knives varied from 24.45 to 26.16 kj/min, respectively. The values of per cent maximum aerobic capacity (VO_2 maximum) and work pulse for sugarcane harvesting knives were much higher than that of the acceptable workload (AWL) limits of 35%. Based on the analysis of results, the sugarcane harvesting knife (H1) ranked as I in terms of minimum value of heart rate (132.55 beats min-1), energy cost of work (24.45 KJ/min), acceptable work load (58.14%), overall discomfort rate (moderate discomfort) and body part discomfort.

Kankal and Desmukh (2014) conducted a study on ergonomic evaluation of manual citrus fruit harvester. The study found that the average work pulse value was 38 beats/min. The average oxygen consumption was found 0.46 l/min. Overall discomfort rating was observed as 2.5, which came under the category of "light" discomfort. Body part discomfort score was found 4.5. The workload by harvester came under "moderate" category. The average output of the harvester was found 2.61 kg/h.

Ghany *et al.* (2015) studied that prickly pear (Cactus) was one of the important plants for sustainable development at arid region. Enhancing picking process and reducing microbial spoilage incident during hand picking, a fundamental goal to face obstacles was spread planting. Due to fruits and cladodes restrictive properties (physical and mechanical), an ergonomic pruning shear was developed to pick fruits. Results showed that using developed aid had a significant effect ($P < 0.05$) on labour productivity and picking cost compared with hand picking. Picking by developed aid increased productivity about 7.2% and decreased cost about 0.04 kg.

Conclusively, the perusal of the literature in this chapter revealed the following:

Adverse environmental conditions badly affect the physiological conditions of workers. Grapes cultivation, vertical farming in which workers were adopting repetitive, strenuous, awkward posture with over stretching and bending posture to accomplish the task. High level of MSDs was prevalent in workers. The available ergo solution was not suitable for the grape workers of Haryana. No detailed study was done on the Occupational risk factor in grapes cultivation and its ergo solution. So, the present study entitled "Occupational risk factor in grape cultivation and its ergo solution" was taken up to fill the gap of the earlier researches, with an aim to assess the work profile and working environment along with occupational risk factor in grapes cultivation and its Ergo solution.

□□□

Chapter 3

Materials and Methods

Research methodology is the foundation for carrying out any research programme. This provides the outline of the methods and procedures to be used during research work. The present research was conducted in three phases. The description of these has been presented under following subheads:

Phase-I: Survey work – A survey was conducted to study the work profile and working conditions of workers in grape cultivation and grape processing practices used by them.

Phase- II: Experiment work - Experiment work was done to study occupational risks of workers in grapes cultivation.

Phase-III: Ergo solution – Ergo solution was developed to overcome the hazards in the highly risks prone activity in grapes cultivation.

The research procedure followed has been described under the following subheads:

3.1 Locale of the study

3.2 Sampling procedure

3.3 Variables and their measurement

3.4 Tools and technique of data collection

3.5 Experimental procedure

3.6 Analysis of data

3.1 Locale of the study

Phase I: Three districts of Haryana state were selected purposively in consultation with Horticultural Department of Haryana state, where grapes cultivation was being done and six grape orchards from these districts were selected randomly for the field study.

Phase II: Two Grape orchards were selected randomly for field experiment from the six grape orchards selected in phase I.

Phase III: Lab of Department of Family Resource Management, I.C. College of Home Science, CCS Haryana Agricultural University, Hisar was used for development of ergo solution and one grape orchard was selected randomly from orchards selected in phase II, for feedback of ergo solution.

3.2 Sampling procedure

Phase I: All the workers working in six grape orchards were selected purposively from each orchard. The work profile of the workers was examined on the basis of type of work, time spent, rest period, body part involved, discomfort rating, pain symptoms, workload as per time and environmental condition in all the activities of grapes cultivation viz land preparation, pruning, manuring, irrigation, plant protection, harvesting and handling and transportation.

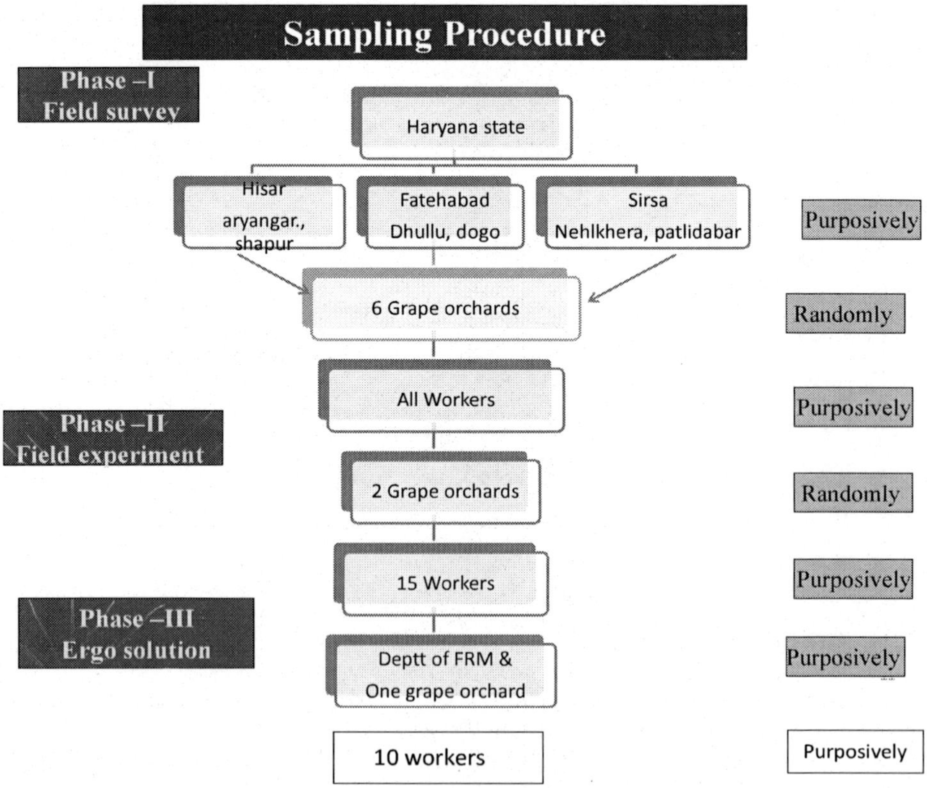

Fig. 3.1: Sampling design

Phase II: A sample of 15 respondents was selected purposively from the randomly selected 2 grape orchards. Out of the six grape orchards selected in phase I Respondents, who were physically fit and willing to co-operate and engaged in grape cultivation activity were selected.

Physical fitness of the workers involved in grapes cultivation activity was ascertained by measuring the four physiological parameters i.e. blood pressure, body temperature, pulse rate and maximum aerobic capacity (Vo_2 Max).

Phase III: Ergo solution were developed in department of FRM and for testing and feedback of ergo solution. One orchard was selected randomly from the orchards selected in phase- II. Ten workers of the orchard were selected purposively to obtain the feedback of developed ergo solution.

3.3 Variables and their measurement

This section presents the selection of dependent and independent variables for the study and their measurement.

Independent variable: The independent variable for the study has been presented in Table 3.1

Table 3.1: Variables and their measurement

Variables	Name of the instruments /tools
Socio-personal parameters	
Age	Categories devised
Family type	Categories devised
Family size	Categories devised
Caste	Trivedi (1963)
Family occupation	Categories devised
Work duration(min)	Categories devised
Rest pause(min)	Categories devised
Physical parameters	
Body weight	Bathroom weighing scale
Height	Anthropometer
Body temperature	Clinical thermometer
Body Mass Index	Quetelet's Index

Body composition	Lange's Skin fold calliper
Blood pressure and pulse rate	Sphygmomanometer and stethoscope, Digital Blood pressure apparatus
Body parts involved	AICRP -2013 Score sheet
Pain Symptoms	AICRP- 2013 Score sheet
Work load as per time	AICRP-2013 Score sheet
Environmental parameters	
Temperature	Thermometer
Heat stress	Heat Index, NOAA (2001-02) (National Oceanic and Atmospheric Administration)
Light	Lux meter
Humidity	Hygrometer
Suspended particulate matter	Air quality monitor
Air contaminant(CO_2)	

Bathroom weighing scale

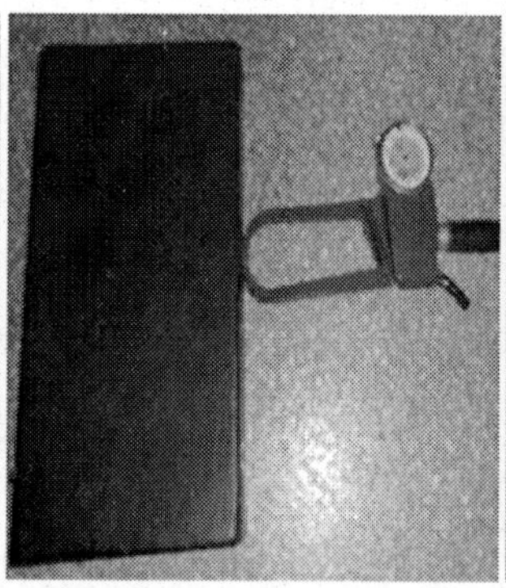

Lange skin fold caliper

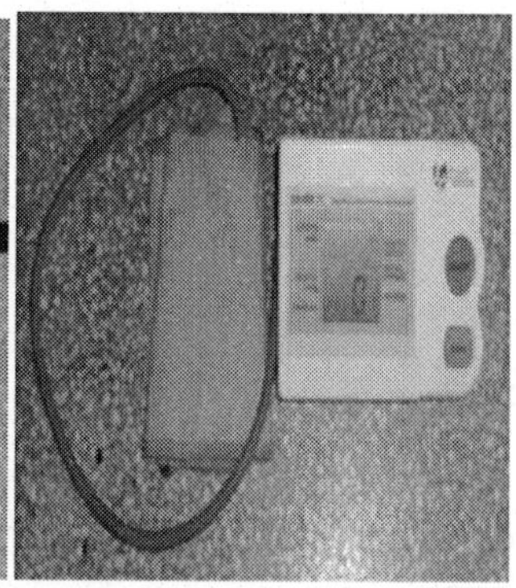

Digital blood pressure

Lux meter

Hygrometer

Air quality monitor

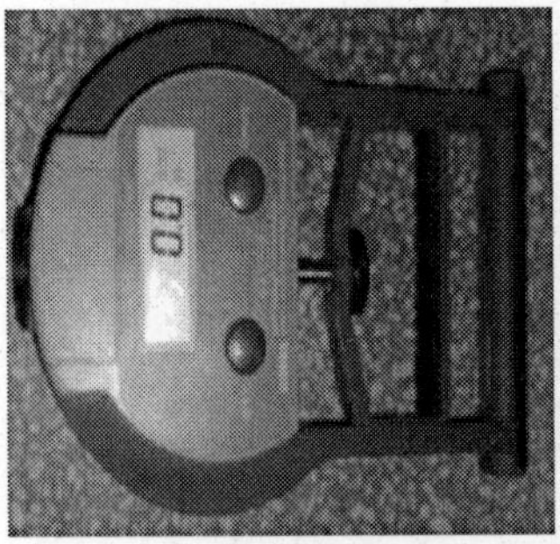

Grip Dynamometer

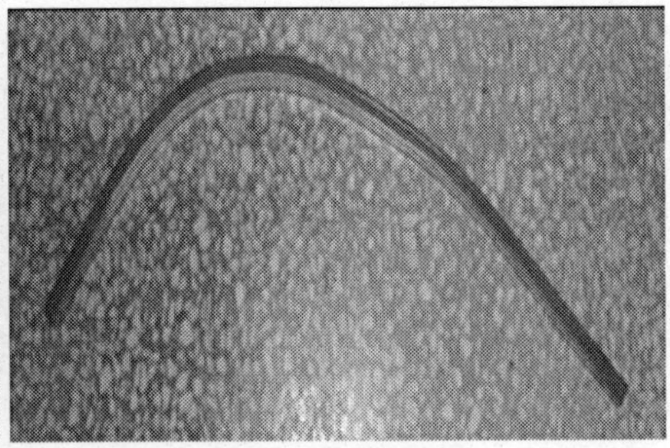

Flexi curve

Polar heart rate monitor

Plate 3.1: Equipments used for the experiment work

Socio-personal parameters

Age: Age was operationally defined as the number of full years completed by grapes orchards workers at the time of interview. The chronological age of the grapes orchards owner & workers at the time of interview was taken by direct questionnaire. For the purpose of analysis, the difference between the lowest and highest age of the respondent was taken and divided by three. On the basis of resultant, following three categories were formed:

- 21-30 years
- 31-40 years
- 41-50 years

Family type: Family type refers to the nuclear or joint type of family. Nuclear family composed of man and a woman joined through marriages and their children living together. Joint family refer to any grouping beyond the nuclear family like husband, wife, their children, also parents and married brothers with their wives and children living together and sharing the common purse. Following categories were made accordingly:

- Nuclear
- Joint

Family size: It refers to total number of members in the family, no matter whether it is nuclear or joint. It was categorized into three groups' i.e small, medium and large. The difference between the maximum and minimum number of members in the family was calculated and then divided by three. The range was made on basis of number obtained by division:

- Upto 3 members (small)
- 4-7 members (medium)
- More than 7 members(large)

Cast: Caste refers to the class or distinct hereditary order of the society and it was categorized as low, medium and high according to modified socio-economic status scale of Trivedi (1963).

- Low : Chamar, Bhangi, Dom, Dhobi, Badi, Jimar
- Medium : Lohar, Kumhar, Dhayal, Baniya, Sonar, Ahir , Saini, Arora
- High : Brahmin, Jat, Rajput, Bishnoi

Working Duration : It includes the working hours spend per day on grapes cultivation activity and was measured in minute.

Rest pause : The period during which physiological function return to resting level is known as the rest pause and measured in minute.

Physical parameters : Physical parameters are related with general health and physical constitution of individual. These are the indicator of physical stamina and capacity to do work. These are directly proportional to the output efficiency and vice-versa with physical structure of workers. Physical parameters used in the study are as follows:

Body weight : It is the vertical force exerted by a mass as a result of gravity. It is an indicator of physical fitness of the person.

Body height: It is vertical distance from the floor to the top of the head, measured while the subject stands erect, looking straight ahead.

Body temperature: Clinical thermometer was used to measure body temperature.

Body Mass Index (BMI):- It is a Key index having relationship of weight to height. BMI was derived by measuring weight and height of the subjects using Quetelet's Index by the following formula given by Garrow (1981):

$$\text{Quetelet's Index} = \frac{\text{Weight (kg)}}{\text{Height}^2 \text{ (m)}}$$

Table 3.2: Grading of health status on the basis of BMI

Scores	Presumption Diagnosis
16.0	CED* grade- III (Severe)
16.0-17.0	CED grade- II (moderate)
17.0-18.5	CED grade-I (mild)
18.5-20.5	Low weight normal
20.5-22.5	Normal
25.0-30.0	Obese grade-I
30.0	Obese grade- II

CED= Chronic energy deficiency

Body composition: It refers primarily to the distribution of muscles and fat in the body and its measurement play a very important role in determining health status of individuals. Body composition was estimated in terms of ectomorph, mesomorph and endomorph population using the scores of Quetelets Index.

Table 3.3: Interpretation of body type on the basis of BMI

Body Type	Quetelet's Index Score	Description
Ectomorph	20	Slender, very thin body
Mesomorph	20-25	Athletic type body
Endomorph	25	Abdominal physical type

Fat weight: Body fat content was measured by using a specially designed skin fold callipers.

Lean body weight: Lean body mass (LBM) was derived by subtracting fat weight from body weight. Estimation of body fat by skin folds measurements.

- **Biceps** skin fold was most bulky portion of the upper front inside arm of the respondents.
- **Triceps** skin fold was most bulky front of the upper backside arm of the respondents.
- **Subscapular** skin fold was just below the bulging portion of subscapular bone of the respondents.
- **Superilliac** skin fold was just above the iliac bone.

To calculate body fat and lean body mass, the following formulae was used:

Sum of skin folds =

Log of skin folds =

Body density (D) = 1.599-(0.0717 × log of sum of skin folds)

$$\text{Per cent fat} = \frac{(4.95 - 4.5) \times 100}{D}$$

$$\text{Fat weight} = \frac{\text{Body weight} \times \%\ \text{Fat}}{100}$$

Lean body mass (Kg) = body weight - fat weight

Blood pressure: Blood pressure is the force of the blood pushing against the walls of the arteries. In the responses to increased demands of the muscles during exercise, the systolic pressure rise more than the diastolic, producing an increase in pulse pressure.The range of normal blood pressure for an adult is 120/80 mmHg; 120 refer to the pressure in the vessels during the heart rate contraction and is called systolic blood pressure and 80 refer to the pressure in the vessels wall, when the heart is relaxing and is referred to as diastolic blood pressure. The pulse pressure is the difference between the systolic and diastolic pressure.

Pulse pressure = Systolic pressure- Diastolic pressure

Average mean pressure = Diastolic pressure + 1/3rd of pulse pressure

Blood pressure of the workers in grapes cultivation activity was taken before the activity and after performing the activity for 2 hours.

Body parts involved: AICRP 2013 Score sheet was used to assess the body parts involved in grapes cultivation

Pain symptoms: AICRP 2013 score sheet was used to assess the pain symptoms in the various body parts during the grapes cultivation.

Workload as per time: AICRP 2013 score sheet was assessed through the workload as per time in grapes cultivation.

Environmental parameters

Temperature: It is the degree of hotness or coldness of the atmosphere. For work, the comfortable range of atmospheric temperature is taken to be about 20 to 25 degree celsius with an average of 23°C. Temperature was measured with the use of Thermometer.

Heat stress: The heat stress index provides general guidelines for assessing potential severity of heat stress. It was recommended that stress can occur at lower temperature that indicated on chart.

Procedure to find on heat stress

- Across the top - locate the relative humidity
- Down the left side of chart, locate the environment temperature
- Follow across down to find out the apparent temperature

Apparent temperature is the combination index of heat and humidity. It is an index of body sensation of heat caused by temperature and humidity.

HEAT INDEX

ENVIRONMENTAL TEMPERATURE (F°)

	70°	75°	80°	85°	90°	95°	100°	105°	110°	115°	120°
Relative Humidity	**Apparent Temperature***										
0%	64°	69°	73°	78°	83°	87°	91°	95°	99°	103°	107°
10%	65°	70°	75°	80°	85°	90°	95°	100°	105°	111°	116°
20%	66°	72°	77°	82°	87°	93°	99°	105°	112°	120°	130°
30%	67°	73°	78°	84°	90°	96°	104°	113°	123°	135°	148°
40%	68°	74°	79°	86°	93°	101°	110°	123°	137°	151°	
50%	69°	75°	81°	88°	96°	107°	120°	135°	150°		
60%	70°	76°	82°	90°	100°	114°	132°	149°			
70%	70°	77°	85°	93°	106°	124°	144°				
80%	71°	78°	86°	97°	113°	136°					
90%	71°	79°	88°	102°	122°						
100%	72°	80°	91°	108°							

*Combined index of heat and humidity...what it "feels like" to the body.

APPARENT TEMPERATURE	HEAT STRESS RISK WITH PHYSICAL ACTIVITY AND/OR PROLONGED EXPOSURE
90° - 105°	Heat cramps or heat exhaustion possible
105° - 130°	Heat cramps or heat exhaustion likely, Heatstroke possible
130° and up	Heatstroke highly likely

Fig. 3.2: Heat Index

Light: It refers to the general lighting conditions in the enterprises in which workers work. Light intensity was measured with the help of Lux meter and the unit of measurement of light is Lux.

Relative humidity: It is an index of amount of water vapours in the air. It is simply percentage saturation of air, which is less than or equal to 100. Relative humidity of 40-50 per cent makes one comfortable in winters while in summers 40-60 per cent is normal. It was measured with the help of Hygrometer.

Suspended particulate matter & Air contamination: Dust concentration were measured with the help of Air Quality Monitor.

II Dependent variable

Dependent variable for the study was occupational risks, as it was assessed through following parameters and has been presented in Table 3.4:

1. Physical load
2. Biomechanical stress
3. Physiological load

Table 3.4: Dependent variables and their measurement

Drudgery parameter	Name of instrument/Formula
1. Physical load	
Load Rating	Weight of the load (5 point continuum scale)
Workload assessment	KIM(Federal Institute for Occupational Safety and Health,2012)
2. Biomechanical stress	
Grip strength (Kg)	Grip dynamometer
Posture	Flexi Curve, Inclinometer and Goniometer RULA score (McAtamney, Corlett.1993) OWAS (Karhu *et al.*1977)
Repetitive Strain (exertion)	ART tool (assessment of repetitive task) Health and safety Executive(2010)Strain index (7-point continuum) (Moore and Garg, 1995)
Musculo-skeletal disorders/discomforts	Human Body Map (Corlett and Bishop,1976)VAD Scale (ODS) (Corlett and Bishop, 1976)NORDIC Scale, (Kuorinka, *et al.* 1987)
3. Physiological load	
Heart rate	Polar heart rate monitor
Oxygen consumption rate (l/min.)	0.0155x HR- 1.2248 (Singh *et al.*, 2007)
Energy expenditure KJ/min	0.159x HR (bpm)-8.72 (Varghese *et al.*1994)
Perceived exertion	RPE Scale (Varghese *et al.*, 1994)

Physical load : Parameter of physical load were:

1. Load rating
2. Work load assessment

Load rating was assessed in grapes cultivation as per Table 3.5

Table 3.5: Load rating in grapes cultivation

Item	Score	Rating
Load rating	Very heavy	5
	Heavy	4
	Moderately heavy	3
	Low	2
	Very Low	1

Workload Assessment

Workload assessment was done through Key Indicator Method (KIM) as per fig 3.3.

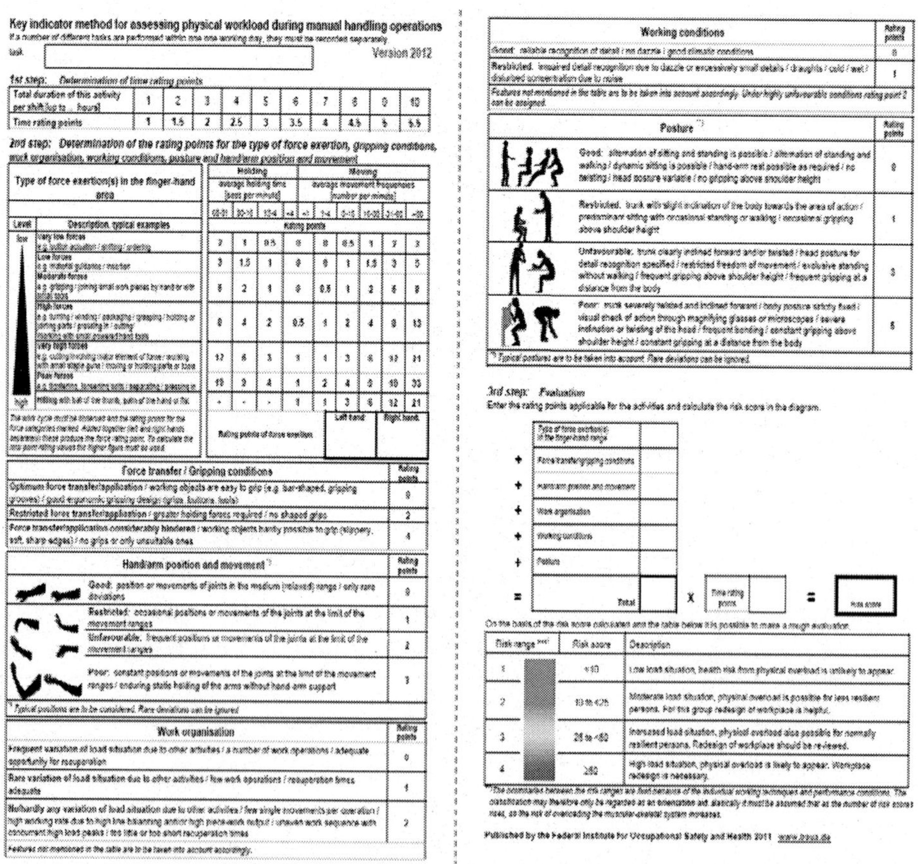
Key indicator method for assessing physical workload during manual handling operations

Version 2012

1st step: Determination of time rating points

2nd step: Determination of the rating points for the type of force exertion, gripping conditions, work organisation, working conditions, posture and hand/arm position and movement

Force transfer / Gripping conditions

Hand/arm position and movement

Work organisation

Working conditions

Posture

3rd step: Evaluation

Fig. 3.3: Workload assessment was done through Key Indicator Method (KIM)

Biomechanical Stress

The stress on the musculo-skeletal system of the workers while workers in grapes cultivation activity was assessed in terms of grip fatigue, postural deviation of spine using following tools and techniques:

Grip strength : Grip strength is the indicator of strength of grip muscles. Reduced grip strength during activity is the indicator of fatigue of that activity. Hence, decreased grip strength while the performance of the activity is interpreted as grip fatigue of the muscles. Grip fatigue is the stress experienced by the grip muscles during or after an activity. It was measured using grip dynamometer. Grip strength of workers was measured before the start of the activity separately for right and left hand and also after the completion of the activity. The grip fatigue was calculated using the following formula:

Grip fatigue (%) = Sr-Sw × 100/Sr

Where,

Sr = strength of muscles at rest

Sw = strength of muscles at work

Reduced muscular strength during activity is an indicator of muscular stress because of the activity.

Postural discomfort : Postural discomforts results, when there is change in normal posture. The postural discomfort was assessed through following scales/tools:

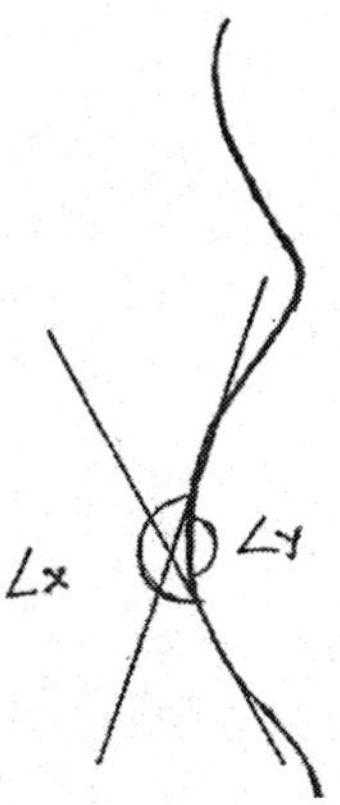

Fig. 3.4: Identificaton Spina angle

Flexi curve : It was used to measure the angle of deviation (degree) between the normal spine curve and maximum deviated posture. The flexi curve was mounted to the contour of the spine and immediately drawn on a plain paper to measure the angle of deviation. The angle of bent of the back during the performance of the activity was measured and compared with the normal bent of the back and the angle of the deviation was determined by subtracting the normal angle of bent from the angle of bent during the bending posture Fig 3.4.

Required angle = x

Measured angle = y

Required angle (x) = 360-"y

In addition to flexi curve, the postural deviation was measured using following scales.

RULA : RULA stands for Rapid upper limb assessment method given by Atamney & Corlett, 1993. It is a survey method developed for ergonomics investigations of workplaces, where work related upper limb disorders are reported. It uses the diagram of body posture including movement of arms, wrist, neck, trunk and legs by scoring method including three scoring table to evaluate the level of exposure of risk factors. A coding system is used to generate an action list which indicates the level of interventions required to reduce the risk of injury due to physical loading on the operator. On the basis of observation of workers in different activities of grapes cultivation, the RULA observation sheet was filled and then grape cultivation activities were numbered to suggest corrective actions and necessary changes as given in Table 3.6 and Fig 3.5

Table 3.6: RULA action sheet

Action	Score	Interpretation
1	1-2	Posture is acceptable, if it is not maintained or repeated for long time
2	3-4	Further investigation is needed and change may be required
3	5-6	Investigation and changes are required soon
4	7-8	Investigation and changes are required immediately

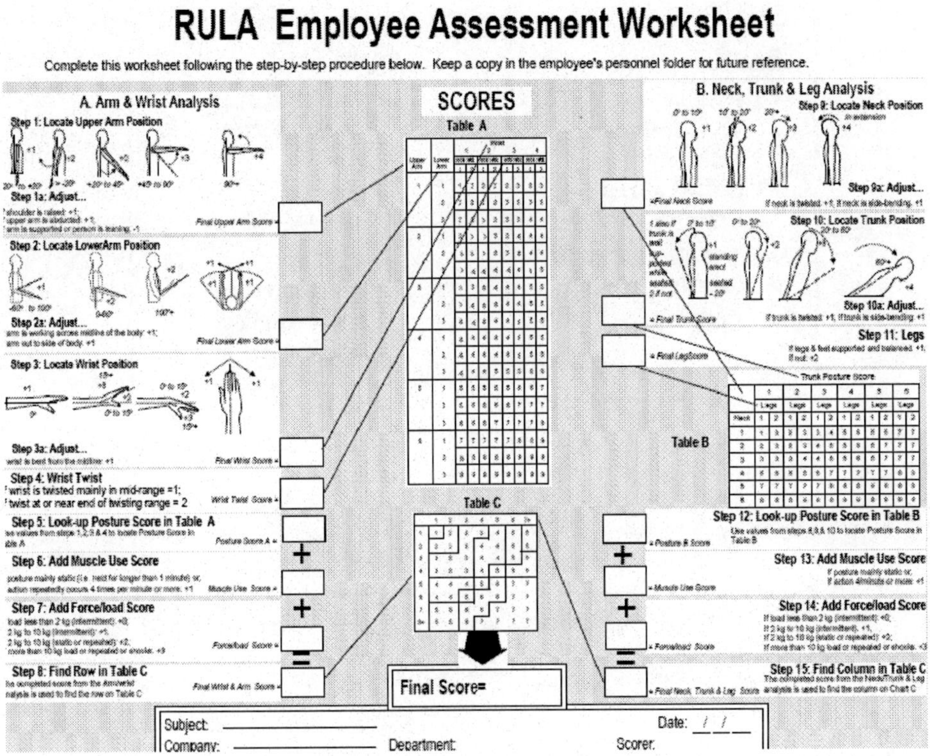

Fig 3.5 : RULA score sheet

OWAS : OWAS stands for Ovako working posture analysis given by Karhu *et al.*, 1977. It is a simple observational method for postural analysis especially that of back, lower and upper extremities and load. Workers in different activities of grape cultivation.The score of OWAS sheet were recorded. The OWAS method uses the concept of number to represent posture with an associated coding system. The jobs with the involvement of high risk were numbered higher and those with the less risk involvement were numbered low and thereafter, immediate corrective actions and necessary changes were recommended as per Fig 3.6 and Table 3.7.

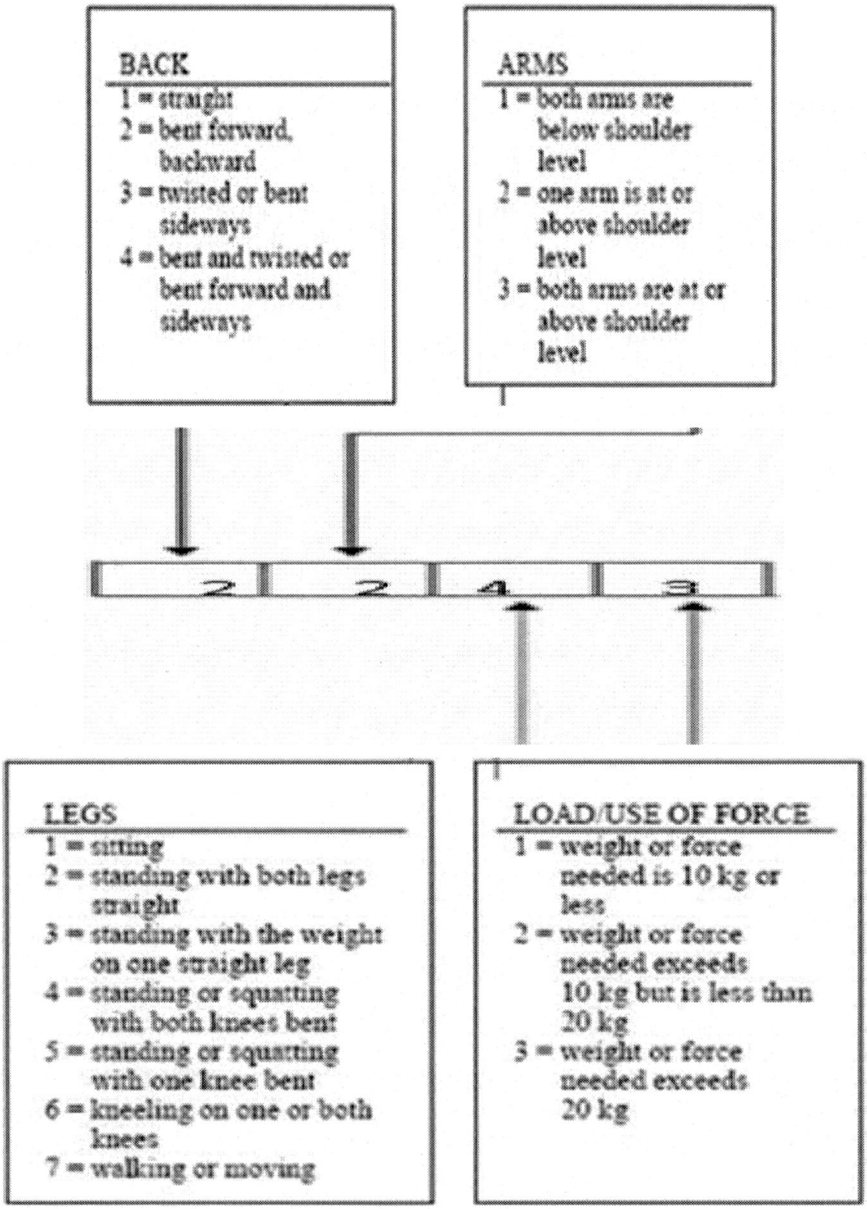

Fig. 3.6 : OWAS score sheet

Table 3.7: OWAS action sheet

OWAS Category	Description
Action category 1	Work postures are usually considered with no particular harmful effects on musculo-skeletal system. No action is needed to change work posture.
Action category 2	Work postures have some harmful effects on musculo-skeletal system. Light stress, no immediate action is necessary, but changes should be considered in future planning.
Action category 3	Work postures have distinctly harmful effects on musculoskeletal system. The working method involved should be changed as soon as possible.
Action category 4	Work postures with extremely harmful effects on musculo-skeletal system. Immediate solutions should be found to change this posture.

Repetitive starin (Exertion) :- Repetative strain or exertion was assessed with the help of ART Tool & Strain Index as per Fig. 3.7.

ART Tool (Assessment of repetitive Tasks of the upper limbs)

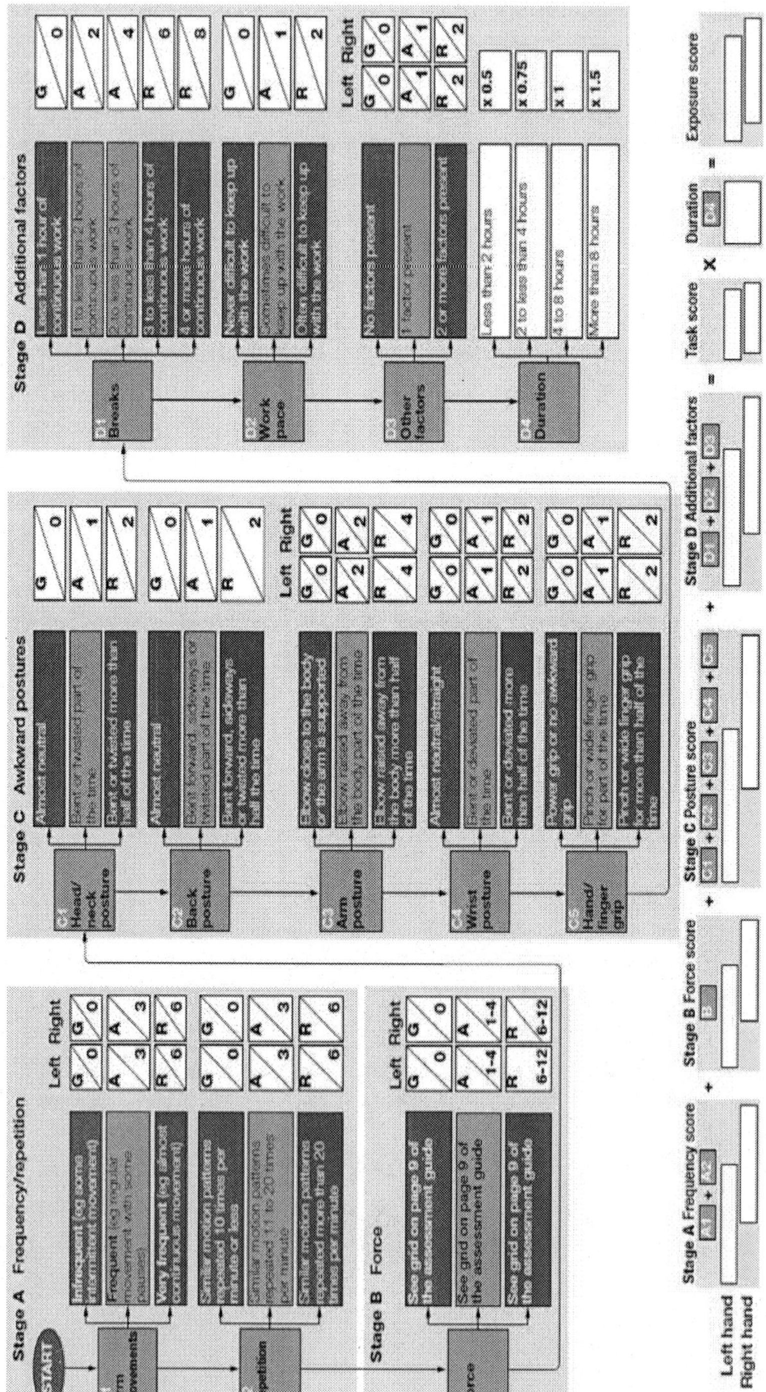

Fig. 3.7 : ART (tool)

Strain index :- Strain index were assessed jobs for risks of work-related musculo-skeletal disorders (WRMSDs) of the distal upper extremities (hand, wrist, elbow) as per Fig 3.8.

Strain Index Scoring Sheet

Date:		Task:	
Company:		Supervisor:	
Dept:		Evaluator:	

Risk Factor	Rating Criterion	Observation			Multiplier	Left	Right
Intensity of Exertion (Borg Scale - BS)	Light	Barely noticeable or relaxed effort (BS: 0-2)			1		
	Somewhat Hard	Noticeable or definite effort (BS: 3)			3		
	Hard	Obvious effort; Unchanged facial expression (BS: 4-5)			6		
	Very Hard	Substantial effort; Changes expression (BS: 6-7)			9		
	Near Maximal	Uses shoulder or trunk for force (BS: 8-10)			13		
Duration of Exertion (% of Cycle)	< 10%	**Calculated Duration of Exertion** (from inputs below)			0.5		
	10-29%	**User Inputs**	**Left**	**Right**	1.0		
	30-49%	Total observation time (sec.)			1.5		
	50-79%	Single exertion time (sec.)			2.0		
	≥ 80%	Number of exertions during observation time			3.0		
	Calculated Duration of Exertion (%)						
Efforts Per Minute	< 4	**Calculated Efforts Per Minute** (from inputs above)			0.5		
	4 - 8		**Left**	**Right**	1.0		
	9 - 14				1.5		
	15 - 19				2.0		
	≥ 20				3.0		
Hand/Wrist Posture	Very Good	Perfectly Neutral			1.0		
	Good	Near Neutral			1.0		
	Fair	Non-Neutral			1.5		
	Bad	Marked Deviation			2.0		
	Very Bad	Near Extreme			3.0		
Speed of Work	Very Slow	Extremely relaxed pace			1.0		
	Slow	Taking one's own time			1.0		
	Fair	Normal speed of motion			1.0		
	Fast	Rushed, but able to keep up			1.5		
	Very Fast	Rushed and barely/unable to keep up			2.0		
Duration of Task Per Day (hours)	<1				0.25		
	1 < 2				0.50		
	2 < 4				0.75		
	4 ≤ 8				1.00		
	> 8				1.50		
Results Key		SI ≤ 3	Job is probably safe				
		3 < SI < 7	Job may place individual at increased risk for distal upper extremity disorders				
		7 ≤ SI	Job is probably hazardous				

Fig. 3.8: Strain Index

Musculo-skeletal disorder/discomfort: Musculo-skeleton disorder were assesed through:

- Human body map (corlett & Bishop, 1976)
- VAD Scale (Corlett and Bishop, 1976)
- Nordic Scale, (Kuorinka, *et al.*, 1987)

Human Body map

It is used to measure the localized discomfort, musculo-skeletal discomfort and intensity of pain in different body parts resulting from postural discomfort. Body part discomfort score (BPDS) is obtained using human body map given by Corlett and Bishop, 1976 fig 3.9.. In this technique, the body is divided into a number of regions. After performing the activity, the subjects were asked to indicate body parts that were most painful. After noting these, the next painful parts were asked and continued till the workers indicate no further pain In this scale, scores were assigned at 5-point continuum ranging from 1-5 i.e. very light exertion (1), light exertion (2), moderate heavy exertion (3), heavy exertion (4), and very heavy exertion (5)

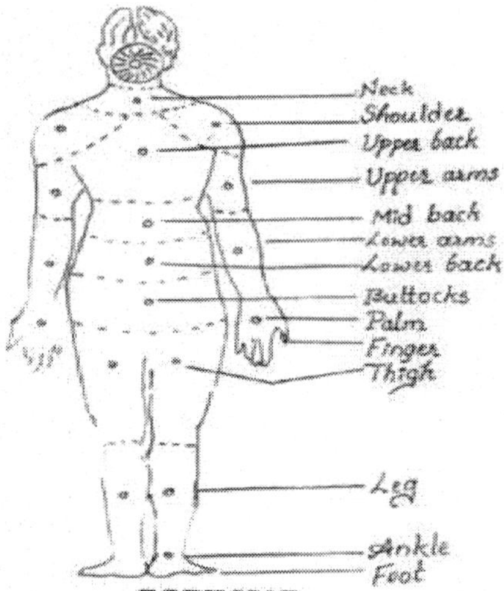

Fig 3.9: Human body map

Visual analogue discomfort scale

For the assessment of overall discomfort rating, a psycho-physical rating scale which is an adaptation of Corlett and Bishop technique, 1976 was used. It is used to determine the discomfort during the activity and doesn't attempt to measure the severity and intensity of the pain. It is a 10-point scale, 0 being the lowest point showing no discomfort and 10 being the uppermost point showing the extreme discomfort. The weighted mean score was derived to reach the conclusion.

Table 3.8: Visual analogue discomfort scale

0	1	2	3	4	5	6	7	8	9	10
No Discomfort										Extreme Discomfort

Nordic musculo-skeletal questionnaire

Nordic musculo-skeletal questionnaire fig 3.10 developed by Kuorinka et al., 1987 was used to gather information on current pain in immediate past 7 days and previous pain in last 12 months. The questionnaire consists of series of objective type questions with multiple type responses. The face to face interview was done to gather required information, as it was thought to be more reliable in obtaining accurate information.

'e you at any time during the **last 12** nths had **trouble (such as ache,** n, **discomfort, numbness)** in:	Have you had **trouble** during the **last 7 days:**	During the **last 12 months** have y **been prevented** from carrying ou normal activities (eg. job, housew hobbies) because of this trouble:
Neck No 1 ☐ Yes 2 ☐	**2 *Neck*** No 1 ☐ Yes 2 ☐	**3 *Neck*** No 1 ☐ Yes 2 ☐
Shoulders No 1 ☐ Yes 2 ☐ in the right shoulder 3 ☐ in the left shoulder 4 ☐ in both shoulders	**5 *Shoulders*** No 1 ☐ Yes 2 ☐ in the right shoulder 3 ☐ in the left shoulder 4 ☐ in both shoulders	**6 *Shoulders (both/either)*** No 1 ☐ Yes 2 ☐
Elbows No 1 ☐ Yes 2 ☐ in the right elbow 3 ☐ in the left elbow 4 ☐ in both elbows	**8 *Elbows*** No 1 ☐ Yes 2 ☐ in the right elbow 3 ☐ in the left elbow 4 ☐ in both elbows	**9 *Elbows (both/either)*** No 1 ☐ Yes 2 ☐
Wrists/hands No 1 ☐ Yes 2 ☐ in the right wrist/hand 3 ☐ in the left wrist/hand 4 ☐ in both wrists/hands	**11 *Wrists/hands*** No 1 ☐ Yes 2 ☐ in the right wrist/hand 3 ☐ in the left wrist/hand 4 ☐ in both wrists/hands	**12 *Wrists/hands (both/eithe*** No 1 ☐ Yes 2 ☐
Upper back No 1 ☐ Yes 2 ☐	**14 *Upper back*** No 1 ☐ Yes 2 ☐	**15 *Upper back*** No 1 ☐ Yes 2 ☐
Lower back (small of the back) No 1 ☐ Yes 2 ☐	**17 *Lower back*** No 1 ☐ Yes 2 ☐	**18 *Lower back*** No 1 ☐ Yes 2 ☐
One or both hips/thighs/buttocks No 1 ☐ Yes 2 ☐	**20 *Hips/thighs/buttocks*** No 1 ☐ Yes 2 ☐	**21 *Hips/thighs/buttocks*** No 1 ☐ Yes 2 ☐
One or both knees No 1 ☐ Yes 2 ☐	**23 *Knees*** No 1 ☐ Yes 2 ☐	**24 *Knees*** No 1 ☐ Yes 2 ☐
One or both ankles/feet No 1 ☐ Yes 2 ☐	**26 *Ankles/feet*** No 1 ☐ Yes 2 ☐	**27 *Ankles/feet*** No 1 ☐ Yes 2 ☐

Fig 3.10: NORDIC Scale

Physiological load

Physiological Load was assessed on the basis of:

- Heart rate
- Oxygen consumption rate
- Energy expenditure
- Perceived excertion

In order to evaluate physiological load (Physiological cost of work PCW), Physiological reaction both during the work and during recovery period are required. A complete work cycle includes physiological cost of work plus physiological cost recovery, which was calculated through heart rate.

Heart rate (HR)

Heart rate of the workers was measured with the help of polar heart rate monitor at rest, during the period of the activity after every minuute and recovery thereafter. Values of resting, average and recovery heart rate were averaged out each time to make a final assessment.

From the values of heart rate Total Cardiac Cost of Work (TCWW) and Physiological Cost of Work (PCW) were calculated using their formula.

TCWW : Total cardiac cost of work

CCW : Cardiac cost of work

CCR : Cardiac cost of recovery

CCW- is the total number of heart beats spent above the resting level in order to perform the work.

CCW= HR× Duration of activity

Where

HR (Increment in HR) = Avg. Working – Avg. Resting HR

CCR = (Avg. Recovery HR- Avg Resting HR) × Duration of activity

CCR- is the total number of heart beats above the resting level occurring between the end of the work and return to the pre- activity state.

TCCW = is the sum total of the CCW and CCR

TCCW = CCW + CCR

PCW = TCCW/ Total time of the activity

The average scores of all the 15 workers were computed using mean and standard deviation and were recorded to obtain the final values.

Oxygen consumption rate : The term 'oxygen consumption' means the amount of oxygen consumed by whole body per unit. It was measured by using formula of Singh *et al.* (2007):

Oxygen consumption, (1/min) = 0.0155 x HR – 1.2248

Energy expenditure : Energy expenditure during work was calculated from the values of average heart rate (AHR) by using the regression equation given by Varghese *et al.* (1994),

Energy expenditure (KJ/min) = 0.159 × Avg working HR (bpm) - 8.72

For classificationm of workload also, the table given by Varghese *et al.* (1994) was used.

Table 3.9: Classiication of workload as per Varghese *et al.* (1994)

Physiological work load	Variables of Physiological cost	
	Energy expenditure (Kj/min)	**Heart rate(bpm)**
Very light	Up to 5.0	Up to 90
Light	5.1-7.5	91-105
Moderately Heavy	7.6-10.0	106-120
Heavy	10.1-12.5	121-135
Very Heavy	12.6-15.0	136-150
Extremely Heavy	≥15.0	≥150

Rating of perceived exertion (RPE scale)

The RPE scale given by Varghese *et al.* 1995 was used to measure the intensity of exertion. It is subjective expression of feelings of workers towards the activity *i.e.* how easy or difficult the subject finds his activity. It is based on physical sensation a person experiences during an activity including increased heart rate, increased sweating and muscle fatigue. This scale now has been accepted as a practical method for rapid appraisal of all occupational work. In this scale, scores were assigned at 5-point continuum ranging from 1-5 *i.e.* very light exertion (1), light exertion (2), moderate heavy exertion (3), heavy exertion (4), and very heavy exertion (5).

Table 3.10 : Rating of Perceived Exertion scale

1	2	3	4	5
Very light exertion	Light exertion	Moderat heavy exertion	Heavy e exertion	Very heavy exertion

3.4 Tools and techniques of data collection

A well-structured interview schedule was developed containing information about number, variety of grape, training system, years of grapes cultivation, mode of payment, training of people, etc.

Phase I : Researcher visited different district Horticulture offices of Haryana state to collect details of grapes orchards including their number and area under grapes orchards. The data was collected form respective district through personnel interview and electronic communication on the basis of this orchard was selected.

In the grape orchards, data were collected personally using pre-tested interview schedule regarding work profile, working conditions of the worker in different activities of grapes cultivation viz. land preparation, pruning, manuring, irrigation, harvesting, trimming and packing and knowledge about grape processing.An observation sheet was used to record the environmental conditions of grapes orchards.

Phase II : An ergonomic study was conducted in field and an observation sheet was used to collect data for occupational risk of workers in grapes cultivation for following parameters:

- Load Rating(kg)
- Workload assessment
- Grip strength(kg)
- Posture
- Repetitive Strain (exertion)
- Musculo-skeletal disorders/discomforts
- Heart rate(bpm)
- Oxygen consumption rate (l/min.)
- Energy expenditure(KJ/min)
- Perceived exertion

Environmental parameters (Temperature, heat stress, light, Carbon dioxide, SPM.) were taken during all the months of year. Data was taken during working hours and three replications of each were taken.

Phase III : A schedule was prepared to get the feedback regarding the appearances, usefulness, ease of use and maintenance of improves grape pruning secateurs and ergo solution). The respondents were asked to reply on each item under each item, the responses were obtained under five-point rating scale. The correct response were scored two and wrong response were given one scores. Mean score for each statement was obtained by multiplying it with respective score obtained and then, dividing it by number of respondents .On the basis of total score, ranking was done.

3.5 Experiment Procedure

Phase I : There was no experiment in Phase I

Phase II : Ergonomics experiments were conducted for occupational risk factor using following procedure:

Physical fitness of the respondents was ascertained by measuring the following physiological parameters as depicted in Table 3.11.

Table 3.11: Parameters of physical fitness

Parameters	Recommended measurement
Blood Pressure	120/80±10mmHg
Body Temperature for 3 minutes	Not above 99 degree F
Heart Rate	60-80 beats per minute
Vo_2(Max)	31.0-40.0 (ml/kg x min)

Assessment of Health status : The health status of the grapes cultivation workers was assessed on the basis of maximum aerobic capacity. Physical fitness is the term which denotes an individual ability to acccomplish a given task in a given period of time aerobically with maximum utilization of oxygen possible. It is also known as maximum aerobic capacity and abbreviated as Vo_2 Max. The maximum aerobic capacity is considered as the best measures for the individual cardio respiratory fitness or capability of doing work. The formula given by Varghese *et.al* (1994) to measure / estimate oxygen consumption was used. The formula was based on the relationship between age and body weight, as they have great influence on VO_2 max.

$(VO_2\,max) = VO_2\,(1/min) = 0.023 \times$ body weight (Kg) $- 0.034 \times$ age (year) $+ 1.652$

$(VO_2\,max)\,(ml/kg \times min) = VO_2\,max\,(1/min/$ Body weight (Kg.)$\times$ 10000

Table 3.12 : Intrepretation of VO_2 scores to determine the health status

VO_2 max (ml/kg × min)	Health status
Up to 15.0	Poor
16.0-25.0	Low average
26.0- 30.0	High average
31.0-40.0	Good
41.0-45.0	Very good
Beyond 45.0	Excellent

The subjects falling within the normal range of health status were selected for the study. The workers performed various activities and all the variables were measured before and after the activity. For each experiment three replications were done.

Phase III

Development and Feasibility testing of Ergo stool for reducing the hazards in the highly risk prone activity in grape cultivation.

On the basis of results of phase II, pruning and harvesting of the grapes were identified as the highly risk prone activities. So, available technology used by them was modified and ergo solution was developed for these two activities.

An ergonomic experiment was carried out on 10 respondents, selected in phase II to find the feasibility of modified and developed ergo solution. Reduction in physiological and biomechanical stress was also assessment in use of modified and developed ergo solution. This phase was carried out in four sub-stages described as follows:

- Modification of existing technology
- Design and development of ergo solution
- Feasibility testing of modified & developed ergo solution
- Physiological and biomechanical assessment of modified and developed ergo solution

Modifications in existing grape pruning secateurs

The secateur were used widely for the pruning and harvesting of grapes by the workers. The use of secateur for long time resulted in grip fatigue and pain mainly in wrist, arm, palm and finger as reported by the respondents in Phase-II.

To overcome these problems, the existing secateur were modified in two ways:

- Covering the handle / hand grip of scateur with plastic pipe
- Covering the handle / hand grip of scateur with raxin cloth

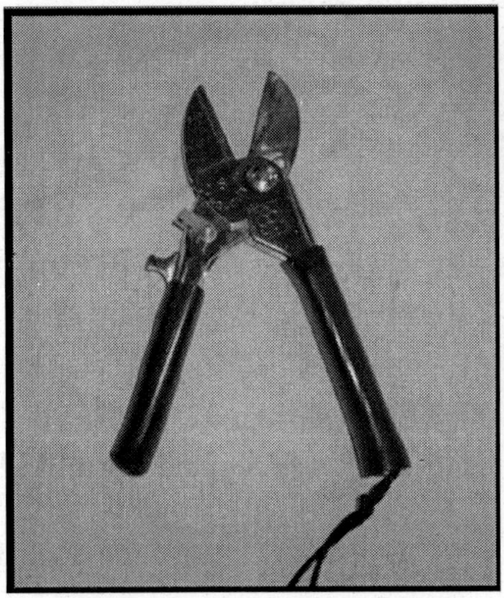

Plastic pipe covered scateur

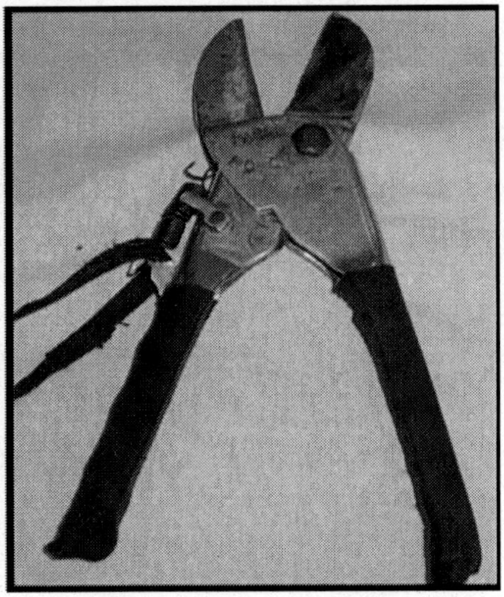

Raxin cloth covered scateur

Plate 3.2: Modifications in existing grape pruning secateurs

Physiological and biomechanical assessment and feasibility testing of both the modified secateurs was done. A schedule was developed to the study feasibility of modified secateurs. Physiological and biomechanical assessment was done through the parameters of biomechanical and physiological load i.e heart rate, oxygen consumption rate, energy expenditure, posture, and grip strength.

Design and development of Ergo stool for grape pruning and harvesting activity

An ergonomically designed sit-stand multipurpose stool was developed considering the anthropometric dimensions of the users populations and it was named as Ergo stool for pruning and harvesting

Design consideration of the Ergo stool for pruning and harvesting.

- **Use of alternate posture** -To allow alternate posture during the activity, an appropriate approach is sit and stand stool. Ergo solution for pruning and harvesting was designed that user can move left and right easily.

- **Storage supplies** - All the supplies, tools should be within the easy reach of workers hence to be provided on the work surface. The drawers were provided in the (Ergo solution for pruning and harvesting for the storage of supplies.
- **Easy to use** - Handle was provided so that it can be easily moved from one place to another.

Ergo solution for pruning and harvesting consisted of platform made up of wood. Two drawer were provided to store items, such as water bottle, scaff, scateurs, blade, scissor, etc. The leg of Ergo solution for pruning and harvesting were simple, strong and made up of steel. On three sides of stool, steel support was provided 5" from the floor level this helped to step on the stool. Handles made up of steel helped to step on the stool.

Design specification of developed (Ergo stool for pruning and harvesting)

- Height - 22"
- Width - 24 "
- Length- 18"
- Handle length- 18"
- Length of rack for storing tool accessories-21"
- Step height - 5"
- Drawers

  i) L×W×H=17×16×6

  ii) L×W×H=16×4×5

(Ergo stool for pruning and harvesting)

Plate 3.3: Design specification of developed Ergo stool for pruning and harvesting

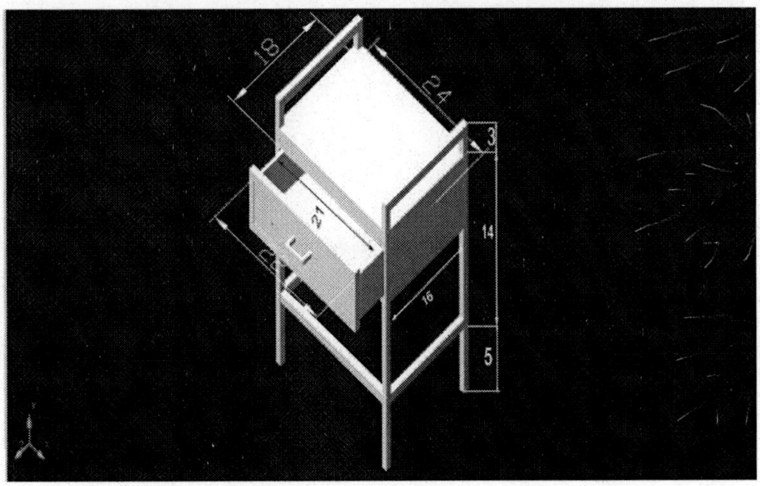

Plate 3.4: Ergo stool for pruning and harvesting

In the present context, feasibility has been defined as the extent to which workers have considered the (Ergo stool for pruning and harvesting and grape pruning secateurs) in terms of different attributes, namely i.e profitability, physical compatibility, cultural compatibility, simplicity, triability, grip fatigue, physical stress factor, work output, tool factor and acceptability. The responses of the workers were recorded on 5 point scale. The quantifying scores for positive statements were quantified by assigning scores i.e.:

- Strongly agree-5
- Agree-4
- Undecided -3
- Disagree-2
- Strongly disagree-1

For quantifying negative statements the scoring was as follows:

- Strongly agree-1
- Agree-2
- Undecided-3
- Disagree-4
- Strongly disagree-5

The mean scores were calculated for each category of statements and attained score were calculated by summation of the mean scores of different statements under each heading. The percentage of maximum attainable score was calculated by using the following formula:

$$\text{Feasibility index:} \frac{\text{E (P+Pc+Cc+ Sc+ Tc+ Gf+ PF+ Wo+ Tf+ Ac)}}{\text{P(P+Pc+Cc+ Sc+ Tc+ Gf+ PF+ Wo+ Tf+ Ac}} \times 100$$

Where,

E (P+Pc+Cc+ Sc+ Tc+ Gf+ PF+ Wo+ Tf+ Ac) is excepted score of profitability, physical compatible, cultural compatible, simplicity, triability, grip fatigue, Physical stress factor, work output, tool factor, acceptability.

P(P+Pc+Cc+ Sc+ Tc+ Gf+ PF+ Wo+ Tf+ Ac is perceived score of profitability, physical compatible, cultural compatible, simplicity, triability, grip fatigue, physical stress factor, work output, tool factor, acceptability.

Physiological and biomechanical assessment of modified tool Ergo stool for pruning and harvesting

Locale : The experiment of modified & (Ergo stool for pruning and harvesting) was done in One grape orchard, selected in phase II of the study.

Selection of respondents : A sample of 10 respondents was selected purposively, from the selected orchard, who were physically fit and willing to co-operate.

Procedure : The workers used the (Ergo stool for pruning and harvesting and grape pruning secateurs) for both activities for one week and Physiological and biomechanical assessment was done as per the procedure mentioned in phase II. Feasibility testing was done through checklist.

3.6 Analysis of data

The responses of the grapes orchards workers were presented, using various statistical techniques like frequency, percentage and mean, weighted mean scores, standard deviation and mean ranks. The physiological and biomechanical parameters were analyzed using mean and standard deviation. Correlation test was applied to find out the association of various variables.

❑❑❑

Chapter 4

Results

This chapter deals with the results of the present research, in accordance with the objectives, inferred through the use of prescribed methodology and standard tools. Results have been presented under the following heads:

4.1 Phase I : Work profile and working conditions of workers in grapes cultivation

4.2 Phase II : Occupational risks of workers in grapes cultivation

4.3 Phase III: Development of an ergo- solution for highly risk prone activity in grapes cultivation

4.1 Work profile and working conditions of workers in grapes cultivation

Working conditions and work profile of grape orchards workers were studied through field survey and walk through investigation of six grape orchards. The background profile of the grape orchard workers have been exhibited in table 4.1 and fig 4.1

Background profile of the workers of grape orchard

Age - Age is very important variable in any study, as it affects the working capacity of an individual. Majority of the respondents (59.3%) in pooled sample belonged to old age and similar trend was also observed in three district i.e majority 50%, 75% and 66.6%, respectively from Hisar, Sirsa and Fatehabad were in old age group. Out of the total sample surveyed, 34 per cent of the respondents belonged to middle age group, 38.8%, 25% and 33.3% from Hisar, Sirsa and Fatehabad, respectively were middle aged. About one-third of the respondents (6.25%) from pooled data were in young age group.

Gender - District wise data indicated that in Hisar, Sirsa and Fatehabad district cent per cent of respondents were male.

Education - There was lot of variation in the level of education of the respondents in three district i.e Hisar, Sirsa and Fatehabad. In Hisar district, 44.4% obtained education upto primary school, 38.8% upto senior secondary, 11.1% of the respondents could read and write and only 5.5 per cent were graduate. In Sirsa district, half of the respondents were educated upto senior secondary level and 37.5 per cent upto primary level. Few respondents (12.5%) were diploma holder. In Fatehabad district, half of the respondents had their education upto senior secondary school and rest half were educated upto primary level.

Pattern of employment - Finding in table 4.1 illustrate that pattern of employment of majority of the respondents (90.6%) in total sample was temporary. Cent per cent from Fatehabad, 88.8 per cent from Hisar and 87.5% from Sirsa had temporary employment. Only 6.2% respondents in pooled sample were having permanent employment, and employment of 12.5% from Sirsa and 55% from Hisar was permanent. Employment of only few respondents (3.1%) was on daily wages in total sample.

Source of income – Further results in table 4.1 revealed that 68.7% respondents in total sample were doing agriculture and 87.5 per cent from Sirsa, 66.6% from Hisar and 50.0% from Fatehabad were involved in agriculture. Whereas 31.3% respondents in total sample, half of the respondents from Fatehabad, one-third from Hisar and 12.5% from Sirsa were involved in private job.

Family type - Regarding family type, the results in table 4.1 unfold that a large number of the respondents (65.6%) in total sample were having joint family, and 55.5 per cent from Hisar, 75.0 per cent and 83.3 per cent from Sirsa and Fatehabad, respectively were having joint family. In total sample 34.3 per cent were having nuclear family and 44.4 per cent from Hisar, 25.0 per cent from Sirsa and 16.6 per cent from Fatehabad, had nuclear family.

Family size – Data in table 4.1 and Fig 4.1 revealed that in pooled sample, 43.7% were having medium family i.e 4-5 members in family, Whereas 50% from both Sirsa and Fatehabad and 38.8% from Hisar were having 4-5 members with medium family. Further Table 4.1 unveiled that 40.6 per cent of the respondents in pooled sample had more than 5 members in family. Fifty per cent from Hisar, 37.5 per cent from Sirsa and 16.6 per cent from Fatehabad had large family. Slighter more than, one-tenth (15.6%) respondents in total sample had small family size i.e. 1-3 members. In Hisar, 11.1 per cent, 12.5 per cent in Sirsa and 33.3 per cent in Fatehabad had small family size.

Marital status - Finding in table 4.1 depicts that majority (84.3%) of the respondents in total sample, all the workers of Sirsa, 83.3% of Fatehabad and 77.7% of Hisar were married. Table 4.1 further depicts that 12.5% workers in pooled sample were single. Whereas, 16.6% respondents, each from Hisar and Fatehabad district were single, none from Sirsa was single. Only few respondents (3.1%) in total sample were widow.

Annual Income - The annual income of the respondents from grape orchards had been presented in table 4.1. Half of the respondents of pooled sample from grape orchards (More than ₹1, 40, 000 /annum) followed by 33.3 per cent having income ranging from (70, 000-1-40000/ annum) and only few (16.6%) were (upto 70,000 per annum). In Hisar, cent per cent had income more than 1,40,000 per annum, In Sirsa (50%), each were in middle i.e 70,001-1,40,000 per annum and more than 1,40,000/annum. In Fatehabad (50%), each were in the medium income group.

Caste - Cent per cent of the respondents in total sample as well as district wise belonged to general category.

Land holding - Most of the respondents (62.5%) in pooled sample had less than 1acres of land, whereas 72.2% from Hisar, and 50% each from Sirsa and Fatehabad had less than 1 acre of land. Landless workers in the total sample were 18.7%, whereas, 25% from Sirsa and 16.6%, each from Hisar and Fatehabad were landless. Overall 9.3% each were marginal and large farmer with 5-10 acres and more 10 acres of land, respectively.

Table 4.1: Background profile of the workers of grape orchard

(n=32)

Variables	Hisar (n=18)	Sirsa (n=8)	Fatehabad (n=6)	Total (N=32)
Age				
Below 18 years (adolscents)	2(11.1)	-	-	2(6.25)
19-25 years (young)	7(38.8)	2(25.0)	2(33.3)	11(34.3)
Above 26 years (late young)	9(50.0)	6(75.0)	4(66.6)	19(59.3)
Gender				
Male	18(100.0)	8(100.0)	6(100.0)	32(100.0)
Education				
Read & write	2(11.1)	-	-	2(6.2)
Primary	8(44.4)	3(37.5)	3(50.0)	14(43.7)
Sen. Secondary	7(38.8)	4(50.0)	3(50.0)	14(43.7)
College	1(5.5)	-	-	1(3.1)
Diploma	-	1(12.5)	-	1(3.1)
Pattern of employment				
Permananent	1(5.5)	1(12.5)	-	2(6.2)
Temporary	16(88.8)	7(87.5)	6(100.0)	29(90.6)
Daily labour	1(5.5)	-	-	1(3.1)
Source of income				
Agriculture	12(66.6)	7(87.5)	3(50.0)	22(68.7)
Private Job	6(33.3)	1(12.5)	3(50.0)	10(31.2)
Family type				
Joint	10(55.5)	6(75.0)	5(83.3)	21(65.6)
Nuclear	8(44.4)	2(25.0)	1(16.6)	11(34.3)
Family size				
1-3 members (small)	2(11.1)	1(12.5)	2(33.3)	5(15.6)
4-5 members(medium)	7(38.8)	4(50.0)	3(50.0)	14(43.7)
Above 5 (large)	9(50.0)	3(37.5)	1(16.6)	13(40.6)
Marital status				
Married	14(77.7)	8(100.0)	5(83.3)	27(84.3)
Single	3(16.6)	-	1(16.6)	4(12.5)
Widowed	1(5.5)	-	-	1(3.1)

Income				
(Upto 70,000) annum	-	-	1(50.0)	1(16.6)
(70,001-140,000) per annum	-	1(50.0)	1(50.0)	2(33.3)
(More than 140,000) per annum	2(100)	1(50.0)	-	3(50.0)
Caste				
General	18(100.00	8(100.0)	6(100.0)	32(100.0)
Land holding				
landless	3(16.6)	2(25.0)	1(16.6)	6(18.7)
Less than 1 acres	13(72.2)	4(50.0)	3(50.0)	20(62.5)
Marginal(5-10)	1(5.5)	-	2(33.3)	3(9.3)
Large(10 acres)	1(5.5)	2(25.0)	-	3(9.3)
Possession of animals				
Cow	-	-	1(16.6)	1(3.1)
Buffalo	18(100.0)	8(100.0)	5(83.3)	31(96.8)

Figures in parenthesis indicate percentage

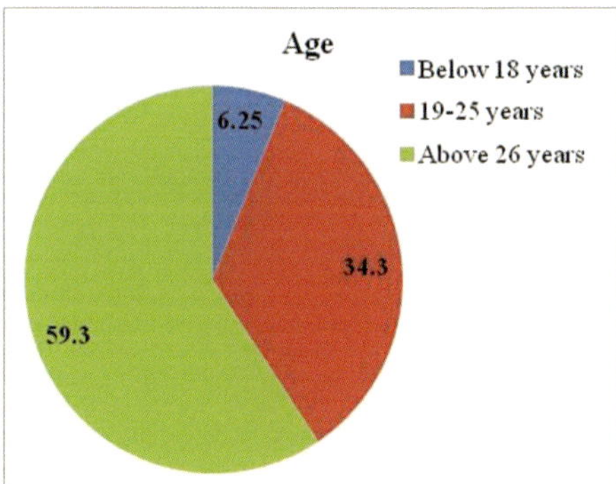

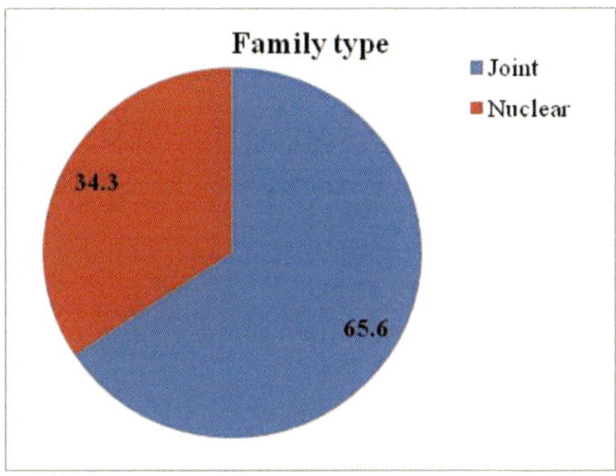

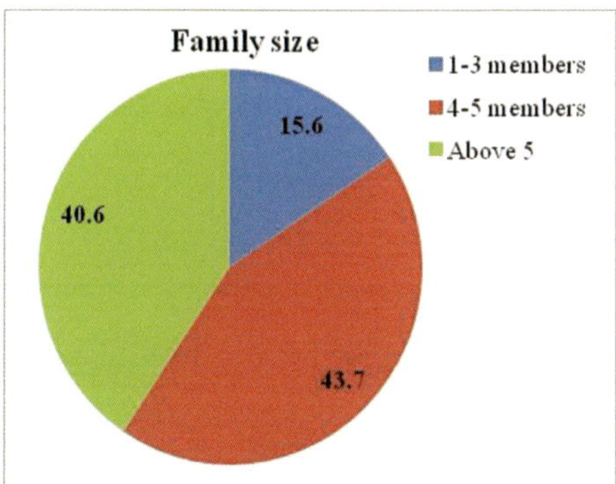

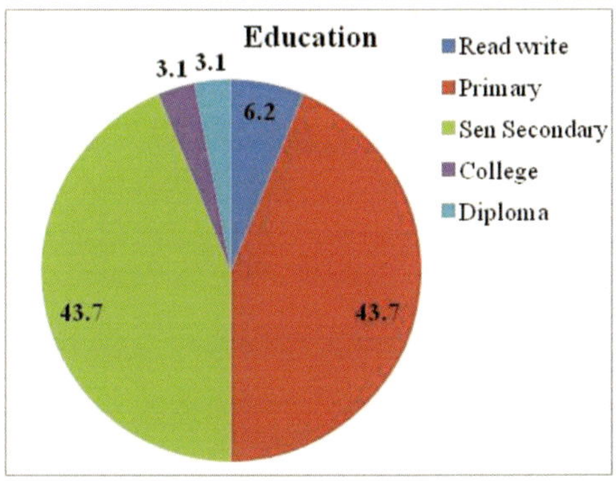

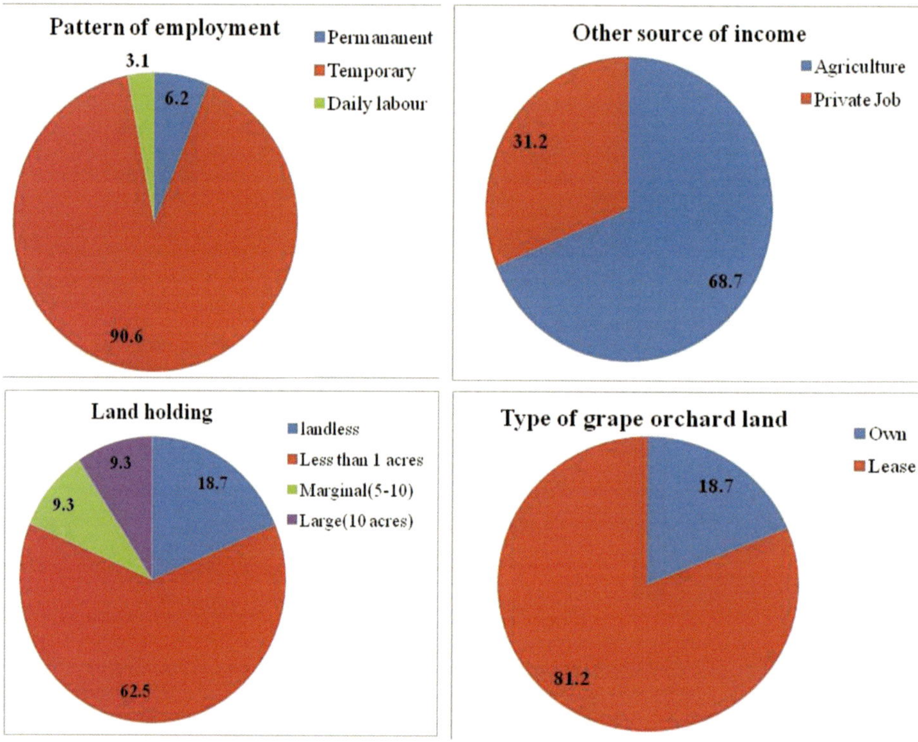

Fig 4.1 : Background profile of the workers of grape orchard

Possession of animals - Finding in Table 4.1, illustrate that majority of the respondents (96.8%) in pooled sample were having buffalo and cent per cent respondents from Hisar and Sirsa, 83.3% from Fatehabad had buffalo. Only few respondents (3.1%) in total sample possessed cow.

Type of land for grapes cultivation

Type of land for grape cultivation and type of grape orchard farming has been presented in Table 4.2.

Type of land - All the respondents of pooled sample had irrigated type of land.

Type of grape orchard farming - The table 4.2 showed that majority of the respondents (81.2%) in total sample had taken the land on lease for grape orchard. District wise land taken on lease for grape orchard by the respondents was 88.8% from Hisar, 75% from Sirsa and 66.6% from Fatehabad. On the other side, 18.7 per cent were having own land for grape orchard. Whereas, 33.3% from Fatehabad, 25.0% from Sirsa and 11.1% from Hisar had own land for grape orchards.

Table 4.2: Type of land for grapes cultivation

(n=32)

	Hisar	Sirsa	Fatehabad	Total
Type of land				
Irrigated	18(100.0)	8(100.0)	6(100.0)	32(100.0)
Type of Grape orchard farming				
Own land	2(11.1)	2(25.0)	2(33.3)	6(18.7)
Lease	16(88.8)	6(75.0)	4(66.6)	26(81.2)

Figures in parenthesis indicate percentage

Years of grape farming - Overall data as well as district wise data indicate that cent per cent of the respondents were doing grape farming from more than 9 years.

Working hours of workers in grapes cultivation

Table 4.3 depicts the working hours in grape orchard. Finding in table 4.3 illustrate that in pooled sample cent per cent of the workers were involved in grapes cultivation from more than 9 years. Overall cent per cent of the respondents were taking rest in between the work, with the length of work period 4-8 hours per days. Results further indicated that majority of respondents (55%) in Hisar district were taking rest for 10 min-60 min and remaining 44.4% were taking rest for 60 min-120 min. In Sirsa, rest period of 10 min-60 min and 60-120 min was taken by 37.5% each, whereas, rest period of more than 120 min was taken by 25.0 % of workers. In Fatehabad, majority (83.3%) were taking rest of 60 min-120 min and only 16.6% were taking rest for of more than 120 min.

Table 4.3: Working hours of workers in grapes cultivation

(n=32)

	Hisar (n=18)	Sirsa (n=8)	Fatehabad (n=6)	Total (n=32)
Years of grape farming				
More than 9 years	18(100.0)	8(100.0)	6(100.0)	32(100.0)
Rest in between the work				
Yes	18(100.0)	8(100.0)	6(100.0)	32(100.0)
Length of work period/day				
4hrs to 8hrs	18(100)	8(100)	6(100)	32(100)
Frequency of rest period				
10min - 60min	10(55.5)	3(37.5)	-	13(40.6)
60min-120min	8(44.4)	3(37.5)	5(83.3)	16(50.0)
More tha120 minutes	-	2(25.0)	1(16.6)	3(9.3)

Figures in parenthesis indicate percentage

Mode of payment of workers in grapes cultivation

The mode of payment of workers of grape orchard has been present in table 4.4. Regarding wages in pooled sample 50% were getting income (wages) weekly, and other 50% were working on contract basis. In Hisar, 55.5% of workers were getting the wages weekly and rest 44.4% of were working on contract basis. In Sirsa, cent per cent worker were contractual labour and in Fatehabad, cent per cent were getting wages weekly. Mode of fixing wages of half of the respondents was based on time of work and that of other half of the respondents was on quantum of work and among these 55.5 per cent were from Hisar and cent per cent were from Fatehabad. Mode of payment of 44.4 per cent workers from Hisar and cent per cent workers from Sirsa was based on quantum of work.

Table 4.4: Mode of payment of workers in grapes cultivation

(n=32)

Mode of payment	Hisar (n=18)	Sirsa (n=8)	Fatehabad (n=6)	Total (n=32)
Weekly	10(55.5)		6(100)	16(50)
Contractual	8(44.4)	8(100)		16(50)
Mode of fixing wage				
Time of work	10(55.5)	-	6(100)	16(50)
Quantum of work	8(44.4)	8(100)	-	16(50)

Figures in parenthesis indicate percentage

Area of grape orchard

Results in table 4.5 depict the area of different grapes orchards. The maximum area of grape orchard was in Hisar district i.e. 45,760 metre square followed by Sirsa 44,000 sq mt and Fatehabad, 39,600 sq mt and least areas were also in Hisar district 36,300sqmt.

Table 4.5: Area of grapes orchard

(n=6)

District	Length(m)	Width(m)	Area of grape orchard(m^2)
Hisar			
1	220	165	36,300
2	220	208	45,760
Sirsa			
1	200	220	44,000
2	190	200	38,000
Fatehabad			
1	180	220	39,600
2	180	220	39,600

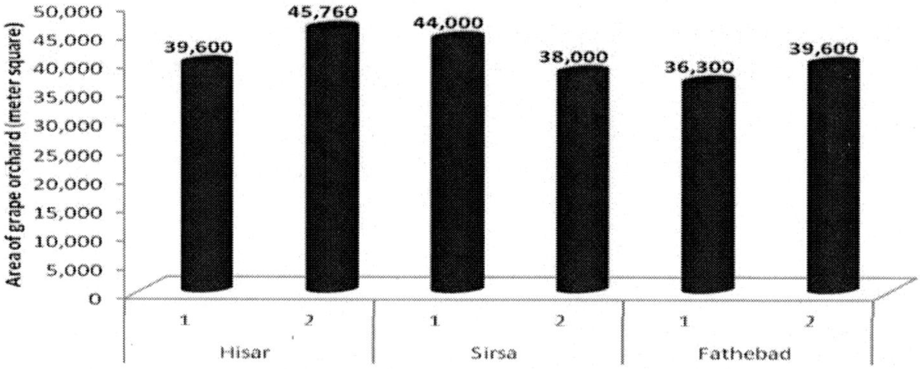

Fig 4.2 : Area of grape orchard

Varieties of grape cultivated

Tables 4.6 represent the varieties of grape cultivated in the grape orchard. In pooled sample, majority of the respondents 66.6% were cultivating perlette variety and 16.6% each were cultivating Anab-e-shabi and Thomson seedless. In Hisar and Fatehabad, perlette variety had been cultivated in all the orchards, In Sirsa 50%, respondents were cultivating Anab-e-shahi and other 50%, respondents were cultivated Thomson variety. On the other side, cent per cent of the respondents were using telephone system for training and knowledge about the raising method for cultivation of grapes.

Table 4.6: Varieties of grape cultivated

(n=6)

Varieties	Hisar	Sirsa	Fatehabad	Total
Anab-e-shahi	-	1(50)	-	1(16.6)
Perletee	2(100)	-	2(100)	4(66.6)
Thomson seedless	-	1(50)	-	1(16.6)
Cultivation in grapes orchards				
Raising	2(100)	2(100)	2(100)	6(100.0)
Training system				
Telephone system	2(100)	2(100)	2(100)	6(100.0)

Figures in parenthesis indicate percentage

Training profile of the owner in orchards

Cent per cent of the owners had acquired training for grapes cultivation. Mode of training of cent per cent was lecture and one-third had attended workshop for grapes cultivation. In Hisar and Fatehabad, all the grape orchard owners had attended the lectures and 50% respondents had attended workshops. In Sirsa, all the cultivators had acquired knowledge through lectures. Results further indicated that overall 83.3% learned about growing grapes cultivation from friend, 16.6% from family. In Hisar, 50% respondents learnt grapes cultivation from family and friends. In Hisar and Fatehabad, cent per cent learnt growing grapes cultivation from family and friends.

Table 4.7: Training profile of the owners in orchards

(n=6)

	Hisar	Sirsa	Fatehabad	Total
Have you taken any training				
Yes	2(33.3)	2(33.3)	2(33.3)	6(100)
Type of training				
Workshop	1(50)	-	1(50)	2(33.3)
Lecture	2(100)	2(100)	2(100)	6(100)
From whom learnt about growing grapes cultivation				
Family	1(50)			1(16.6)
Friend	1(50)	2(100)	2(100)	5(83.3)
Self-experience	-	-	-	

Note: Multiple response
Figures in parenthesis indicate percentage

Knowledge and practices of grape orchard owners about grape processing

Table 4.8 presents the knowledge of owners of grape orchards regarding grape processing. In pooled sample, 50% of owners had knowledge that grape harvested were fit for sale by upto 3 days, 33.3 % felt that it can be sold by upto 2 days and only 16.6% were of view that harvested grapes can be sold by upto one day. In Hisar district, 50% respondents felt that grapes can be sold by upto 2 days of harvesting and remaining 50% were of opinion that it can be sold by upto 3 days.

Knowledge about grape processing : Cent per cent of the owners had knowledge about the grape processing.

Use of unsold grapes : Two-third of owners had knowledge that unsold grape can be used for making raisins. Fifty per cent owners in Hisar district and also in Fatehabad had knowledge that unsold grape can be used for making raisins. Cent per cent owners in Sirsa had knowledge that unsold grape can be used for making raisins. Half of the owners from three districts did not know what to do with unsold grapes. One-third of the owners in pooled sample and half of the owners in both district, Sirsa and Fatehabad were discarding the unsold grapes. Few of the owners (16.6%) in total sample had knowledge that grapes can be used for making wine and other 16.6% felt that grapes can be used for chutney making.

Table 4.8: Knowledge and practices of grape orchard owners about grape processing practicing

(n=6)

Grapes practices	Hisar	Sirsa	Fatehabad	Total
Grapes harvested fit for sale				
Upto 1 day	-	-	1(50)	1(16.6)
Upto 2 day	1(50)		1(50)	2(33.3)
Upto 3 day	1(50)	2(100)		3(50.0)
Knowledge regarding grape processing				
Yes	2(100)	2(100)	2(100)	6(100.0)
Use the unsold grapes				
Raisins	1(50)	2(100)	1(50)	4('66.6)
Wine	-	1(50)	-	1(16.6)
Chutney	1(50)	-	-	1(16.6)
Do not know	1(50)	1(50)	1(50)	3(50.)
Discard	-	1(50)	1(50)	2(33.3)
Use of seedless grapes for making for grape product				
Yes	-	1(50)	-	1(16.6)
No	2(100)	1(50)	2(100)	5(83.3)
Grape products made through processing of grapes				
Raisins	2(100)	2(100)	2(100)	6(100.0)
Jam	2(100)	1(50)	-	3(50.0)

Juices	2(100)	-	-	2(33.3)
Pickle	-	1(50)	1(50)	2(33.3)
Wine	2(100)	1(50)		3(50.0)
Use the grapes leaves for chutney				
Yes	2(100)	2(100)	2(100)	6(100.0)
Use any other part of grape plant for making products				
No	2(100)	2(100)	2(100)	6(100.0)
Grape processing units near orchard				
No	2(100)	2(100)	1(50)	5(83.3)
Yes	-	-	1(50)	1(16.6)

Figures in parenthesis indicate percentage
Multiple responses

Use of seedless grape for making grape products : In total sample, 83.3% of the owners did not had knowledge that seedless grapes were used for making grape products. Cent per cent of owners in both, Hisar and Fatehabad district and half of the owners in Sirsa district did not had knowledge that seedless grapes can be used for making grape products. Only 16.6% owners in pooled sample had knowledge that seedless grapes can be used for making grape products.

Grape products made through processing of grapes : All the owners in the total sample had knowledge that raisins can be made by processing of grapes. Jam and wine can be made through the processing of grapes was know to 50% of the owners and 33.3% knew that grapes can be used to make juices and pickle. Cent per cent of the owners of Hisar district had knowledge that raisin jam, juice and wine can be made through processing of grapes. All the owners of Sirsa had knowledge that raisin can be made and half had knowledge that jam/pickle and wine can be made through grape processing. In Fatehabad, cent per cent knew that raisin can be made and 50% had knowledge that pickle can be made through processing of grapes.

Use the grape leaves for chutney : In pooled sample as well district wise cent per cent had knowledge that grape leaves can be used for making chutney.

Use any other part of grape plant for making products : Cent per cent of the grape orchard owner in pooled sample as well as district wise had no knowledge that other parts of grape plants can be used for making any products.

Grape processing units near orchard : Only 16.6% owners in total sample had knowledge about the grape processing unit near orchard, whereas 83.3% owner did not had knowledge about availability of grape processing unit near the orchard. In Hisar and Sirsa, all the owner and in Fatehabad (50%) of owner did not had knowledge, about the availability of grape processing unit near the orchard.

Level of tiredness and pain faced by workers in grapes cultivation

Data in table 4.9 represents the level of tiredness and pain faced by grape orchard workers. Cent per cent of the respondents in pooled sample and also in district wise sample felt tiredness while doing activity. Further table 4.9 unveiled that majority of the respondents (31.2%) were feeling tired after working for 3-4 hrs/day, followed by 28.1% who felt tiredness after working more than 4hrs, 21.8% were tired after 1-2 hrs and 18.7% felt tiredness after 2-3hrs. In Hisar district, 38.8% felt pain after 3-4 hr, of workers 33.3% after more than 4hrs, 16.6% after 1-2 hrs and 11.1% after 2-3hrs. In Sirsa, equal percentages (25%) of worker were in each category of time period. Whereas, in Fatehabad, 33.3% each felt tiredness after 1-2 hrs and 2-3 hrs of work and 16.6% each felt tiredness after working 3-4 hrs and more than 4hrs.

Feeling body pain and discomfort : In Hisar, 50% felt body pain and discomfort sometimes, (44.4%) felt discomfort always and only (5.5%) felt pain and discomfort frequently. In Sirsa, 50% felt pain and discomfort always followed by 37.5%who felt pain sometimes and 12.5% felt pain and discomfort frequently. In Fatehabad, equal percentage of workers (33.3%) felt body pain sometimes, always and frequently. Results further revealed that in pooled sample, 43.7 per cent each felt body pain and discomfort, sometimes and always and only 12.5% felt pain and discomfort frequently.

When you feel pain : Half of the respondents felt pain while doing work and rest of half of the respondents felt pain after work in total data as well as in three districts.

Table 4.9: Level of tiredness and pain faced by workers in grapes cultivation

(n=32)

Level of tiredness and pain	Hisar (n=18)	Sirsa (n=8)	Fatehabad (n=6)	Total (n=32)
Feeling tired while doing this activity				
Yes	18(100)	8(100)	6(100)	32(100)
After how many hours you start feeling tired				
1-2hrs	3(16.6)	2(25.0)	2(33.3)	7(21.8)
2-3hrs	2(11.1)	2(25.0)	2(33.3)	6(18.7)
3-4hrs	7(38.8)	2(25.0)	1(16.6)	10(31.2)
More than 4hrs	6(33.3)	2(25.0)	1(16.6)	9(28.1)
Feeling body pain and discomfort				
Sometime	9(50.0)	3(37.5)	2(33.3)	14(43.7)
Always	8(44.4)	4(50.0)	2(33.3)	14(43.7)
Frequently	1(5.5)	1(12.5)	2(33.3)	4(12.5)
When you feel pain				
While doing work	9(50.0)	4(50.0)	3(50.0)	16(50.0)
After work	9(50.0)	4(50.0)	3(50.0)	16(50.0)
Treatment taken for pain and discomfort				
Massage	-	3(37.5)	3(50.0)	6(18.7)
Medicine	6(33.3)	4(50.0)	1(16.6)	11(34.3)
Rest	12(66.6)	1(12.5)	2(33.3)	15(46.8)

Figures in parenthesis indicate percentage

Treatment taken for pain and discomfort : In total sample to overcome, the pain and discomfort maximum (46.8%) of workers were taking rest followed by 34.3% who took medicine and only 18.7% had taken massage for it. In Hisar (66.6%), were taking rest and 33.3% were taking medicine to overcome the pain and discomfort. In Sirsa to overcome pain and discomfort, 50% were taking medicine, 37.5% were taking massage and only 12.5% were taking rest. In Fatehabad, 46.8% were taking rest, 34.3% were taking medicine and 18.7% were taking massage to overcome the pain.

Levels of problems faced by workers in grape orchard

Data in Table 4.10 represent the levels of problem in different grape orchards. Major problems which were physical problem, workplace problem, problem due to high temperature, problem related to work, weather and climate, pesticides, economic problems, postural problem and injuries and insects bites. Physical problems include backache, asthma skin problems, pain in legs and pain in hand, numbness in body, swelling in hands, tingling sensation. Data in table further reflect that overall weighted mean score of physical problems of workers in three districts (Hisar, Sirsa, Fatehabad) was 2.19 and rank IV. Workplace problems, include lack of cleanliness and lack of safety measures and overall score was 2.37 with rank III. Problems due to high temperature include burn skin, eye strain, vomiting and headache with overall weighted mean score of 2.12 with rank VII.

Regarding problems related to work included hard physical work, exertion due to overwork, rude behaviour of owner, and workers and abusive language which were observed in Hisar, Sirsa and Fatehabad with overall score 2.13 and rank VI.

Data in table 4.10 showed that problems due to pesticides included sneezing, breathlessness and eye irritation of workers. The problems due to pesticides scored overall weighted mean score of 2.25 and rank V. Economics problems include delay in wages and low wages of workers. Overall score of economic problems was 2.08 with rank VIII. Another problems were postural problems which include pain due to stiffness of joints, pain in toes due to prolong stress and unable to stand properly. The rank I was achieved by postural problems with overall score of 2.67.Problems related to injuries/ insects bites included cut, fracture, wound, fall and insects biting of workers which ranked II with overall score of 2.63.

Table 4.10: Level of problems faced by workers in grape orchard

n=32

Problems	Weighted mean score (x_w)			Mean score	Ranking
	Hisar (n=18)	Sirsa (n=8)	Fatehabad (n=6)		
Physical problems	1.94	2.16	2.48	2.19	IV
Workplace problems	2.3	2.75	2.05	2.37	III
Problems due to high temperature	2.1	2.12	2.15	2.12	VII
Problems related to work	2.12	2.22	2.06	2.13	VI
Problem due to Pesticides	2.26	2.26	2.23	2.25	V
Economics problems	1.95	1.9	2.4	2.08	VIII
Postural problems	2.63	2.76	2.63	2.67	I
Injuries/cuts	2.46	2.7	2.72	2.63	II

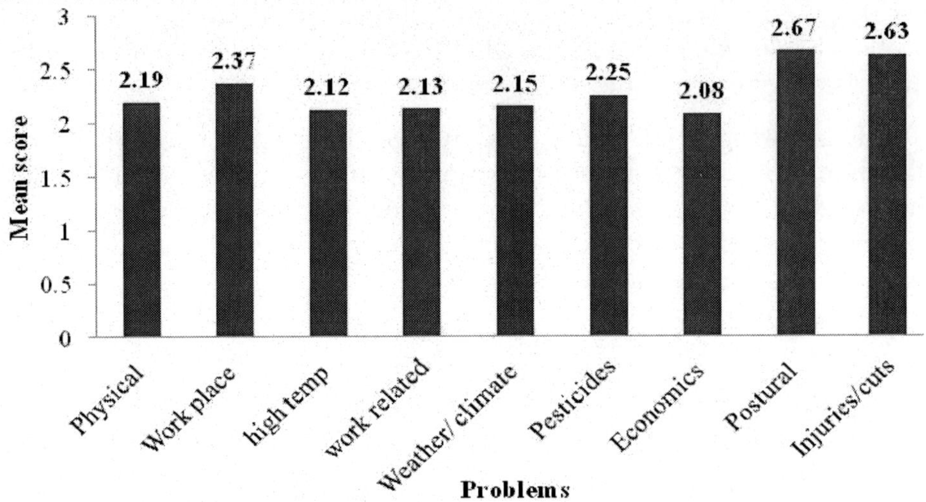

Fig 4.3: Level of problems faced by workers in grape orchard

Postures used and risk prone areas of body in activities performed by the workers in grapes cultivation

The finding in table 4.11 presents the activities undertaken by workers in grapes cultivation along with posture adopted, work schedule, risk prone areas of body, period of activity, number of workers and technology used for each activities.

Land preparation : Land preparation activity was done by 26 workers in different work schedule morning /afternoon/evening during the month of December through use of *spade.* The bending posture was adopted, by workers and risk prone areas of body were lower back, neck, mid back.

Pruning : Pruning was done by 30 workers in different work schedule morning /afternoon/evening during the month of January through use of scateur. The activity was performed by the workers in standing posture with raised arm, and risk prone areas of body were feet, lower back shoulder, upper arms.

Manuring : Manuring activity was done by 8 workers in different work schedule i.e morning /afternoon/evening during the month of Feb to April through tractor and manual spray pumps. The workers were performing activity in standing posture and risk prone area were feet and lower back.

Irrigation : Irrigation activity was done by 6 workers in different work schedules during the month of Feb to April through tube-wells. The workers were in standing posture while performing activity and risk prone areas and were feet lower back.

Harvesting : Harvesting activity was done by 29 workers during the month of May/ June either manually or use of shaving blade. The harvesting was done in standing posture with risk prone areas of feet, lower back and mid back. Packing activity was done manually by 8 workers during the month of May to June. The workers were packing of grapes in sitting without back support, and risk prone areas were lower back and mid back.

Transportation : Transportation activity was done by 8 workers. During the activity, the workers were in standing posture with risk prone areas in lower back, mid back and feet.

Land preparation

Pruning

Irrigation

Manuring

Plant protection

Harvesting

Transportation

Plate 4.1: Activities performed by workers in grape orchards cultivation

Table 4.11: Posture adopted and risk prone areas of body in activities performed by the workers in grapes cultivation

Activities	No. of worker involved	Work schedule	Period of activity	Posture adopted	Risk prone areas of body	Posture	Technology used
Land Preparation	26	Morning/afternoon/ evening	December	Bending	Lower back, Neck, Mid back		Spade
Pruning	30	Morning/Afternoon/ evening	January	Standing with raised arm	Feet , lower back Shoulder , upper arms,		Scateur
Manuring	8	Morning /afternoon/ evening	Feb/March/ April	Standing	Feet, Lower back		Tractor Manually Spray pumps
Irrigation	6	Morning/afternoon/ evening	Feb/March/ April	Standing	Feet , Lower back		Tube wells
Harvesting	29	Morning/afternoon/ evening	May /June	Standing	Feet , lower back,mid back		Manually/ shaving Blade
Transportation	8	Morning/afternoon/ evening	May/June	Standing,	Feet, lower back		Manually

Environmental conditions of grapes orchards

A compatible work environment is very important to perform the task easily and to increase the productivity with reduced numbers of errors. Hence, the grape orchard environment was assessed with environmental parameters which include temperature, relative humidity, light, CO_2 and SPM (suspended particules matters. Table 4.12 revealed the data regarding the environment conditions of grapes orchards in winter season as well as summer season.

Temperature (^{0}C) : Table 4.12 revealed that recommended value of the temperature was 20- 24^{0}C. In winter season, mean temperature in the orchard in Hisar district was 24.1^{0}C, 20.4^{0}C in Sirsa district and by 19.5^{0}C in Fatehabad district. During summer season, mean temperature in the orchard in Hisar district was 31.9^{0}C followed by 29.3^{0}C in Sirsa district and 28.7^{0}C in Fatehabad district.

Relative humidity (%) : Recommended value of the relative humidity is (40-60%).Table 4.12 further revealed that in winter season, mean relative humidity in the orchards in Hisar district was 93.1% and 92.0% in Sirsa district and Fatehabad district. During summer season, mean relative humidity in the orchards was 78.2% Hisar district and 76.7% in Sirsa district and 78.2% in Fatehabad district.

Light (Lux) : Recommended value of the illumination level is 500-1000 lux.Table 4.12 highlighted that the lighting conditions in the orchards, during winters in Hisar district, mean illumination levels in the orchards was 536.8 lux followed by 611 lux in Sirsa district and 665.0 lux in Fatehabad district. During summer season, in Hisar district mean illumination levels in the orchards was 848.8 lux, 611.5 lux in Sirsa district and 898 lux in Fatehabad district.

Carbon dioxide (CO_2ppm) : Recommended value of CO_2 is 350-450ppm Table 4.12 revealed that CO_2 level during winters in the orchards, in Hisar district was 320.4ppm followed by 308.3ppm in Sirsa district and 235.8 in Fatehabad district. During summer season, CO_2 level in the orchards in Hisar distrct was 307.8ppm, 308.3ppm in Sirsa district and 285.9ppm in Fatehabad district.

SPM (Suspended particules matters ppm) : Recommended value of SPM is 140ug/m^3. Table 4.12 revealed that suspended particulate matter during winters in Hisar district was 45.114 ug/m^3 followed by 53.7140ug/m^3 in Sirsa district and 45.1140ug/m^3 in Fatehabad district. During summer season, in Hisar district SPM was 405.1ug/m^3 and 53.7ug/m^3 in Sirsa district and 423ug/m^3 in Fatehabad district.

Table 4.12 : Environmental condition of Grape orchards

Seasons	Hisar		Sirsa		Fatehabad		Re-commen-ded value
	Surround-ing of orchard	In the orchard	Surround-ing of orchard	In the orchard	Surround-ing of orchard	In the orchard	
Temperature (^{0}C)							
Winter	24.1±148.4	21.7±9.50	22.7±12.7	20.4±0.49	21.8±3.5	19.5±3.5	20-24 (^{0}C)
Summer	35.4±5.6	31.9±17.9	34.1±5.6	29.3±15.0	31.9±6.5	28.7±15.8	
Relative humidity (%)							
winter	90.4±11.7	93.1±1.5	89.4±4.0	92.0±6.3	89.4±2.8	92.0±2.8	40-60%
summer	76±11.0	78.2±4.2	74.4±36.6	76.7±4.2	74.4±4.2	78.2±4.2	
Light (Lux)							
Winter	764.4±8.7	536.8±3.6	764.4±4.7	611±5.8	764.4±2.8	665.0±2.8	500-1000 Lux
Summer	1212±4.8	848.8±18.5	764.4±4.7	611.5±5.8	1152.0±4.2 4	898.6±4.2	
Co_2 (ppm)							
Winter	337.2±4.7	320.4±6.5	335.1±4.2	308.3±6.1	291.1±2.8	235.8±.2.8	350-450ppm
Summer	342.1±8.8	307.8±15.6	335.1±4.2	308.3±6.1	324.9±4.2	285.9±4.9	
Spm (µg/m^3)							
Winter	59.3±3.5	45.1±6.8	59.6±4.8	53.7±5.7	51.2±2.8	45.1±2.8	140 ug/m^3
Summer	506.4±10.8	405.1±18.1	59.6±4.8	53.7±5.7	481.1±21.6	423±17.4	

Heat stress faced by workers in grapes cultivation

Heat stress faced by workers in grape cultivation was assessed in winter and summer season and has been presented in table 4.13

Winter season : In winter season, the environmental conditions in grapes orchards of Hisar, Sirsa and Fatehabad was comfortable with heat stress score 79, 71 and 71, respectively. This indicates that there was no problem due to the heat stress in all 3 districts.

Summer season : Table 4.13 further depicts the heat stress in grape orchard during summer season. The heat stress score in grape orchards of Hisar was 113, and that of Sirsa and Fatehabad was 92 and 94, respectively. The score of these 3 district depicts that extreme caution was required as this heat stress results in sun stroke, heat cramps and heat exhaustion possible with prolonged physical activity.

Table 4.13 : Heat stress faced by workers in grapes cultivation

(n=6)

Orchards	Heat scores	Interprpation	Effect of heat stress
		Winter season	
Hisar	79	Less than 80 normal	No problem
Sirsa	71	Less than 80 normal	No problem
Fatehabad	71	Less than 80 normal	No problem
		Summer season	
Hisar	113	90-104 (extreme caution)	Sun stroke , heat cramps and heat exhaustion possible with prolonged physical activity
Sirsa	92	90-104 (extreme caution)	Sun stroke , heat cramps and heat exhaustion possible with prolonged physical activity
Fatehabad	94	90-104 (extreme caution)	Sun stroke , heat cramps and heat exhaustion possible with prolonged physical activity

Phase-II

4.2 Occupational risks of workers in grapes cultivation

The occupational risk in grapes cultivation was assessed through physical load, biomechanical stress, physiological and time load of the respondents. This section includes the physical, physiological, biomechanical and physiological parameters of the respondents involved in grapes cultivation activities. These parameters of the respondents were recorded from the time of start of the activity till its completion.

Physical characteristics of workers involved in grapes cultivation

Table 4.14 depicts that mean height and weight of grape workers involved in grape was 159.9 cm and 64.2 kg, respectively. Body mass Index (BMI) was observed as 21.8 Kg/m^2. Fat percentage was worked out to be 29.9%. Hence, LBM (Lean body mass) was 44.1 kg with variation of ±19.3kg. Aerobic capacity (VO_2 max) was found to be 31.8 ml/kg.×min exhibiting that the subjects were having good health.

Table 4.14: Personal profile and health status of the selected respondents

(n=15)

Physical Characteristics		**Mean ± SD**
Height (cm)		159.9± 8.8
Weight (kg)		64.2±4.7
BMI (kg/m^2)		21.8±1.1
Body composition	Fat percentage (%)	29.9±5.9
	Lean body mass (kg)	44.1±19.3
Vo_2 max (ml/kg.×min)		31.8±6.3

Body type - Majority of the workers (80%) were having mesomorphic (Athletic type body), which is considered as the perfect body type followed by 20 per cent were having ectomorphic (cylindrical type thin body).

Table 4.15: Body type of the selected respondents

(n=15)

Body Type	**Q.I**	**Frequency**	**%**
Ectomorphic (Cylindrical very thin body)	<20	3	20
Mesomorphic (Athletic type body)	20-25	12	80

Health status of the grapes workers of involved in grape cultivation

In order to avoid any experimental error and to maintain the uniformity in data, only those workers were selected for the experimental study who had high average to good health status. It was clear from the table 4.16 that the mean body temperature was 98.6±0.33, blood pressure was systolic(mm/Hg)125.6±10.3, diastolic(mm/Hg) 80.3±4.8, pulse rate was 86.8±9.3bpm and pulse pressure was 45.7±4.2 mm/Hg. All the variables were in the normal range.

Table 4.16: Health status of workers involved in grapes cultivation

(n=15)

Variables of health status	Observed values	Recommended values	Category
Body temperature(^{0}F)	98.68±0.33	98.4degree F	Normal
Blood pressure :			
Systolic(mm/Hg)	125.69±10.36	120mm/Hg	Normal
Diastolic(mm/Hg)	80.38±4.89	80mm/Hg	
Pulse rate(bpm)	86.8±9.36	70-80bpm	Normal
Pulse pressure (mm/Hg)	45.7±4.2	30-50mm/Hg	Normal

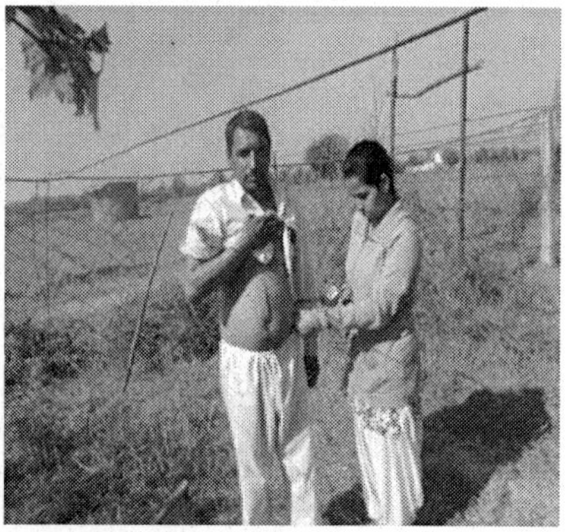

Body composition of the workers

Body weight of the workers

Grip strength of the workers

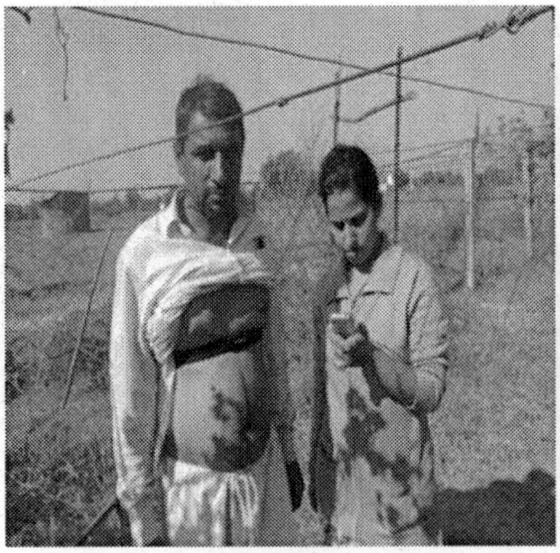

Heart rate of the workers

Angle of deviation of the workers

Blood pressure of the workers

Plate 4.2: Personal profile and health status of the selected respondents

Occupational risks of workers in grapes cultivation

Occupational risks of workers in grapes cultivation were assessed through following parameters :

i) Physical load

ii) Biomechanical stress

iii) Physiological load

iv) Time load

I Occupational risks in grapes cultivation as per physical load

Occupational risk in grapes cultivation as per physical load were assessed through :

- Load rating
- Workload assessment through Key Indicator Method(KIM)
- Physical load

Occupational risks of workers involved in grapes cultivation as per load rating

The scrutiny of table 4.17 reveals that land preparation was very heavy activity with score of 5. Pruning and harvesting got score 4, which means that these were heavy activity. Whereas, handling and transportation activity scored 3, which implies that these were considered as moderately heavy. Manuring, irrigation, plant protection got the score 2, so these were considered as low activity in terms of load rating.

Hence, land preparation was very heavy activity as per load rating.

Table 4.17: Occupational risks of workers involved in grapes cultivation as per load rating

(n=15)

Activity	Load rating	
	Score	**Rating**
Land preparation	Very heavy	5
Pruning	Heavy	4
Manuring	Low	2
Irrigation	Low	2
Plant protection	Low	2
Harvesting	Heavy	4
Handling and transportation	Moderate heavy	3

Load rating: Very heavy-5, Heavy-4, moderately heavy-3, Low-2, Very Low-1

Assessment of physical workload during manual handling operation in grapes cultivation through Key Indicator Method (KIM)

The various aspects under key indicator method included total duration of activity per shift time rating points, type of force exertion in the finger, hand range of both hands, force transfer/gripping conditions, hand arm position and movement, work organization, working condition, posture and evaluation of both hands. The table 4.18 presents the KIM for assessing physical workload in grapes cultivation.

Land preparation: Key indicator method score for both hands was 45, resulting in overall score of 90. The risk score for both hand depicts the increased load situation, physical overload, so re-designing of workplace should be done for land preparation activity.

Pruning : Key indicator method score for left hand was 48.75 and for right hand it was 75 and overall score was 123.75. KIM score for left hand indicate increased load situation, physical overload may possible hence, redesigning of workplace should be reviewed. KIM score for right hand indicate high load situation, physical overload is likely to appear, so workplace re-design is necessary

Manuring : The key indicator method score for left hand was16 and for right hand it was 32, overall score was 48. KIM score of left hand depicts moderate load situation, physical overload also possible for less resilient persons so re-design of workplace should be reviewed. KIM score for right hand indicate increased load situation, physical overload also possible for normally resilient persons. Redesign of workplace should be reviewed.

Irrigation : The key indicator method score for left hand was 17 and 38 for right hand. Overall KIM score was 55. The KIM score of left hand depicts moderate load situation, physical overload also possible for less resilient persons, so re-design of workplace should be reviewed. KIM score for Right hand depicts increased load situation, physical overload also possible for normally resilent persons, so, re-design of workplace should be reviewed.

Plant protection : The key indicator score for both hands was 23, resulting in overall score of 46. The risk score for both hands depicts the increased load situation, physical overload. So, re-design of workplace should be done.

Harvesting : Key indicator method score for left hand and right hand was 70 ,72.5 respectively, and overall score was 142.5.The risk score for both hands depicts the increased load situation, physical overload hence, re-design of workplace should be done.

Table 4.18: Assessment of physical workload during manual handling operation in grapes cultivation through Key Indicator Method (KIM)

(n =15)

	Determination of time rating points		Type of force exertion in the finger hand range		Force transfer/gripping conditions	Hand / arm position and movement	Work organization	Working conditions	Posture	Evaluation of left hand and right hand	Description	Rank
	Total duration of this activity per shift {up to hours}	Time rating points	Left hand	Right hand						Type of force exertion in the finger hand+Forces transfer Hand arm work organization+ working +		
Land preparation	4	2.5	6	6	2	2	2	1	5	Left hand= 6+2+2+2+1+5= 18×2.5=45 Right hand=6+2+2+2+1+5= 18×2.5=45 **Total = 90**	**Both hands-** Increased load situation, physical overload also possible for normally resilient persons. Redesign of workplace should be answered	**IV**
Pruning	4	2.5	5.5	16	4	2	2	1	5	Lefthand= 5.5+4+2+2+1+5 =19.5×2.5=48.75 Right hand=16+4+2+2+1+5 =30×2.5=75 **Total = 123**	**Left hand=** Increased load situation, physical overload also possible for normally resilient persons. Redesign of workplace should be reviewed. **Right hand=** High load situation, physical overload is likely	**II**

Handling and transportations : Key indicator method score for left hand was 50 and 67.5 for right hand. The risk score for both hands highlighted increased load situation, physical overload. Therefore, re-design of workplace should be done.

On the basis of KIM score, right hand in pruning activity and both the hands in harvesting and handling and transportation were highly risk prone, as there was high load situation, physical overload is likely to appear and hence, workplace re-design is necessary.

Physical load of workers in grapes cultivation

Physical load of workers in grapes cultivation was assessed and presented in table 4.19 and fig 4.4

Land preparation : Physical load factor during various activities of land preparation viz the removing of stalks and stubbles and unwanted plants was 5.02 for ploughing by country plough was 5.83. Mean physical load factor for land preparation was 5.42.

Pruning : Pruning activity includes cutting of undersirable vines and physical load factor was 24.

Manuring : During manuring activity transportation of manure got physical load factor of 46.51and mixing of manure got 74, spreading of manure got 21.3. Mean physical load factor of manuring was 47.2.

Irrigation : Irrigation activity includes preparation of irrigation channels and physical load factor was 24.5.

Plant protection : Activities among plant protection were spraying and topping which got physical load factor of 19 and 8, respectively. Mean physical load factor of plant protection was 13.5.

Harvesting : Under harvesting activity fruit picking got physical load factor of 11.5, gathering and heaping got 1, packaging in polythene got 2.25 and trimming got physical load factor of 11.5 and mean physical load factor of harvesting was 6.55.

Handling and transportation : Physical load factor in handling and transportation for loading of the product was 6.55. The physical load in grape cultivation was highest for manuring, followed by irrigation and pruning.

Table 4.19: Physical load of workers in grapes cultivation

(n=15)

Farm Activity	Physical load / acre					Mean physical load factor
	Weight of the load (kgs)	Distance carried (kms)	Height lifted (mts)	Physical Load Rating[1] (score)	Physical Load Factor (Total Score)	
Land preparation						
Removing of stalks and stubbles and unwanted plants plants	2	0.02	1	2	**5.02**	**5.42**
Ploughing	0.75	0.08	2	3	**5.83**	
					Total =10.85	
Pruning						
Cutting of undesirable vines	6	3	5	10	**24**	**24**
					Total =24	
Manuring						
Transporation of manure	40	0.01	1.5	5	**46.51**	**47.2**
Mixing of manure	70	0	1.0	3	**74**	
Spreading of manure	2.8	14	1.5	3	**21.3**	
					Total = 141.8	
Irrigation						
Preparation of irrigation channels	12	7	1.5	4	**24.5**	**24.5**
					Total= 24.5	
Plant protection						
Spraying	2	14	0	3	**19**	**13.5**
Topping	0	7	0	1	**8**	
					Total= 27	
Harvesting						
Fruit picking	4	3	1.5	3	**11.5**	**6.55**
Gathering and heaping	0	0	0	1	**1**	
Packaging in polythene	.25	0	0	2	**2.25**	
Trimming	4	3	1.5	3	**11.5**	
					Total = 26.2	
Handling and transportaion						
Loading of the product	2	0.05	1.5	3	**6.55**	**6.55**
					Total =6.55	

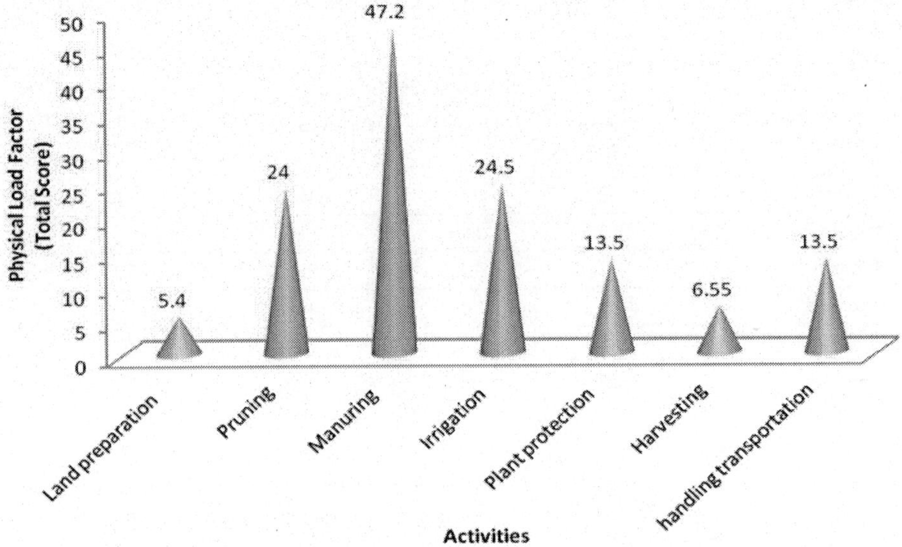

Fig. 4.4: Physical load of workers in grapes cultivation

Occupational risks of grape orchards workers in terms of physical load

The occupational risk of workers in terms of physical load was assessed on the basis of load rating, workload assessment through key indicator method and physical load rating and presented in table 4.20. The physical load was maximum for pruning with total physical load of 178, followed by harvesting having physical load of 153.05. The total physical load of other activities handling and transportation (127.05), land preparation (100.04), manuring (97.2) and irrigation (81.5). The least total physical load was for plant protection (61.5).

On the basis of physical load in the grapes cultivation, pruning activity had highest occupational risk.

Table 4.20: Occupational risks of grape orchards workers in terms of physical load

Activities	Physical load			Total physical load	Rank
	Load rating	KIM	Physical load factor		
Land preparation	5	90	5.42	100.4	IV
Pruning	4	150	24	178	I
Manuring	2	48	47.2	97.2	V
Irrigation	2	55	24.5	81.5	VI
Plant Protection	2	46	13.5	61.5	VII
Harvesting	4	142.5	6.55	153.05	II
Handling and transportation	3	117.5	6.55	127.05	III

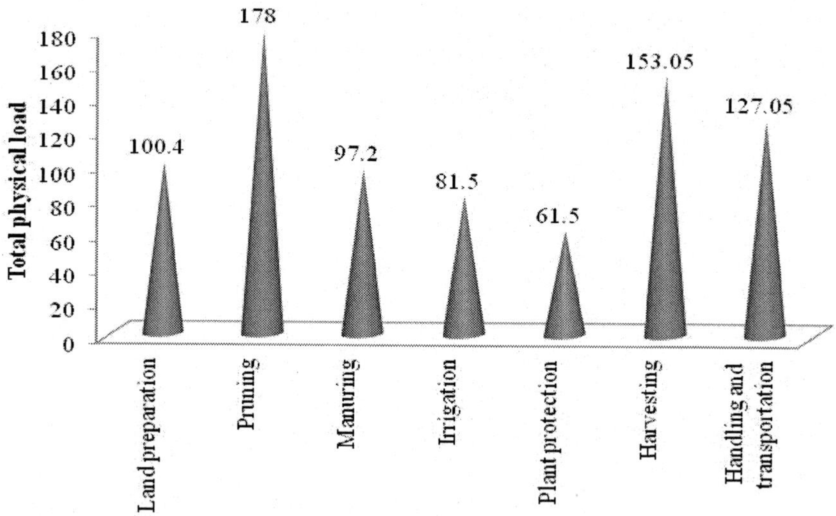

Fig. 4.5: Occupational risks of grape orchards workers in terms of physical load

II Occupational risks in grapes cultivation as per biomechanical stress

Occupational risk in grapes cultivation as per biomechanical stress was assessed through:

- Grip fatigue
- Angle of deviation
- RULA score (Mc Atamney, Corlett. 1993)

- OWAS (Karhu *et.al.* 1997)
- Posture load score
- Repetitive strain exertion
- Musculoskeltal discomfort

Grip fatigue before and after performing the activities in grapes cultivation

Biomechanical stress was assessed by measuring the grip strength of the subjects before and after performing various activities in grape orchard and per cent change was noted down, and presented in Table 4.21 and Fig 4.6.

Land preparation : Average grip strength before and after the land preparation for left hand was 36.0 kg and 25.2 kg respectively. This resulted in the grip fatigue was 30%. For the left hand grip, strength was 31.8 kg and 22.3 kg, respectively, before and after the activity respectively resulting in fatigue of 29.8%.

Pruning : The average grip strength of the right hand before and after activity was 29.7kg and 20.5 kg, respectively, which resulted in grip fatigue was 30.9% similarly for left hand before and after the pruning average grip strength was 25.1 kg and 17.3 kg, respectively. The grip fatigue of the left hand was 31.0%.

Manuring : The average grip strength of the right hand before and after the activity was 26.3 kg and 23.0 kg, respectively, which resulted in grip fatigue was 20%. The average grip strength of the left hand before and after performing the manuring was 22.1 kg and 17.6 kg, respectively. The grip fatigue of left hand was 20.3%.

Irrigation : Before and after performing the irrigation, the average grip strength of the right hand was 20.5 kg and 16.4kg respectively. This resulted in grip fatigue was 20%. Similarly, for left hand it was 18.2kg and 14.6 kg, respectively. The grip fatigue of the left hand was 19.7%.

Plant protection : Before and after performing the plant protection activity, the average grip strength of the right hand was 23.7kg and 19.9 kg, respectively. This resulted in grip fatigue was 15.7%. Similarly, for left hand it was 21.5 and 20.0, respectively. The grip fatigue of the left hand was 6.7%.

Table 4.21: Grip fatigue of worker before and after performing the activities in grapes cultivation

(n=15)

Activity	Grip strength before activity (kg)	Grip strength After activity (kg)	Grip fatigue (%)
Land preparation			
Right hand	[illegible]	25.2±9.98	30
Left hand	31.8±10.8	22.3± 11.3	29.8
Pruning			
Right hand	29.7±11.3	20.5±11.3	30.9
Left hand	25.1±10.3	17.3±10.3	31.0
Manuring			
Right hand	26.3±11.04	21.0±11.3	20.1
Left hand	22.1±10.5	17.6±8.69	20.3
Irrigation			
Right hand	20.5±12.2	16.4±11.5	20
Left hand	18.2±9.73	14.6±8.92	19.7
Plant protection			
Right hand	23.7±12.2	19.96±11.5	15.7
Left hand	21.50±9.73	20.05±8.92	6.7
Harvesting			
Right hand	34.7±10.3	29.5±10.5	14.9
Left hand	31.1±11.0	26.4±10.9	15.1
Handling and transportation			
Right hand	37.7±12.2	32.5±11.5	13.7
Left hand	34.1±9.73	29.4±8.92	14.9

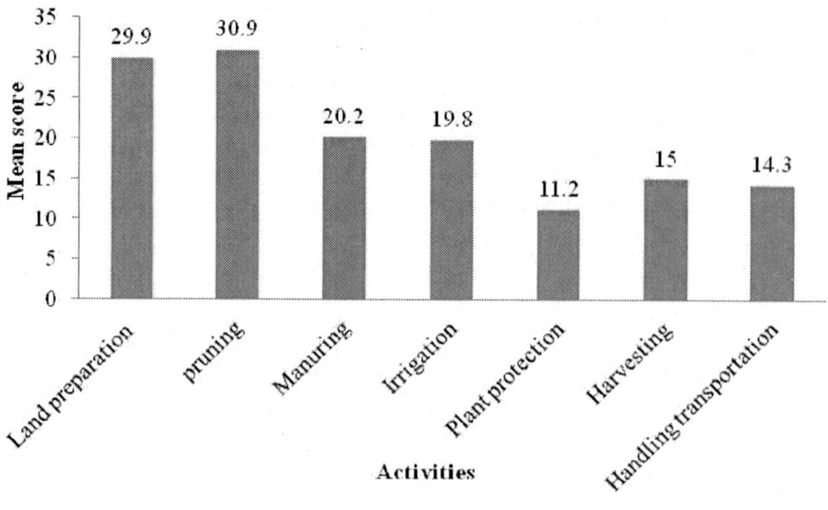

Fig. 4.6 : Grip fatigue of worker before and after performing the activities in grapes cultivation

Harvesting : In harvesting activity, the average grip strength before and after the activity for right hand was 34.7kg and 29.5 kg, respectively. This resulted in grip fatigue of 14.9%. The average grip strength of the left hand was 31.1 kg and 26.4 kg, respectively and grip fatigue of the left hand was 15.1%.

Handling and Transportation : In handling and transportation activity, the average grip strength before and after the activity for right hand 37.7 kg and 32.5, respectively. This resulted in grip fatigue of 13.7%. The average grip strength of the left hand 34.1 kg and 29.4 kg, respectively and grip fatigue of the left hand was 14.9%.

The grip fatigue was maximum in pruning activity for both the hands.

Spinal angle deviation in different activities in grapes cultivation

The finding in Table 4.22 and fig 4.7 highlights the angle of back (spinal cord) of workers in normal position and during work. The normal angle of workers was found to be 181.9 ± 2.8^0. The maximum forwarded deviation in angle was observed in land preparation 17.1, followed by manuring and irrigation, plant protection 7.9 each, pruning 7.1 and least deviation was in harvesting and handling and transportation 5.1. This resulted in maximum percent deviation in spinal angle in land preparation 9.4% followed by manuring and irrigation, plant protection 4.3%, each pruning 3.9% and least deviation was for harvesting, handling and transportation (2.8%).

The maximum deviation in the spinal angle was in land preparation followed by manuring, irrigation, plant protection and pruning and least was for harvesting and handling and transportation.

Table 4.22: Spinal Angle deviation in different activities in grapes cultivation

(n=15)

Activity	Normal (degree)	Working (degree)	Deviation in angle	Per cent deviation %
Land preparation	181.9±2.8	199±1.6	17.1	9.4
Pruning		189±0.7	7.1	3.9
Manuring		174±1.8	7.9	4.3
Irrigation		174±1.8	7.9	4.3
Plant protection		174±1.8	7.9	4.3
Harvesting		187±1.4	5.1	2.8
Handling and transportation		187±1.4	5.1	2.8

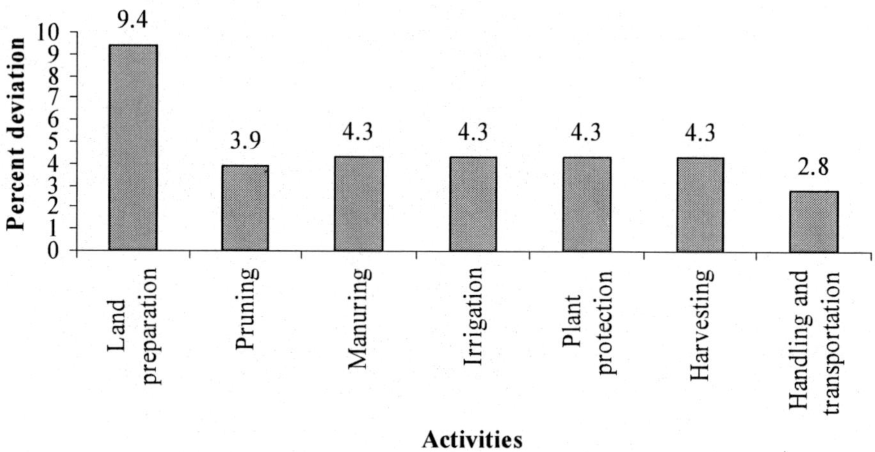

Fig. 4.7: Spinal Angle deviation in different activities in grapes cultivation

Rapid upper limb assessment (RULA) scores of workers in grapes cultivation

Table 4.23 depicts the rapid upper limb assessment (RULA) of posture used during land preparation, harvesting, manuring pruning, irrigation, plant protection and handling and transportation activity of grapes cultivation.

Land preparation : The mean value of wrist and arm score was 3.0 and muscles used forces was 2, thus giving the total score of 5. The mean score for neck, trunk and leg, worked out to be 6 with muscle force of 1, and total score (score B) of 6. Thus, leading to a mean final score of 7. This total scores of 7 generates an action category investigate and change immediately, in the land preparation activity.

Table 4.23: Rapid upper limb assessment (RULA) score of workers in grape cultivation

(n=15)

Posture	Posture Score A (upper arm+ Lower arm+wrist+wrist twist	Muscle used Force	Score A	Posture Score B(Neck, Trunk and leg score)	Muscle used Force	Score B	Final score	Action category	Description	Rank
Land preparation	2+2+3+1=3	1+1	5	3+4+1=5	1+0	6	7	7	Investigate and change immediately	I
Pruning	5+3+3+2=7	1+0	8	5+4+1=8	1+0	9	+7	7	Investigate and change immediately	I
Manuring	3+3+2+1=4	0+0	4	3+4+1=5	0+0	5	5	5	Investigate further and change soon	III
Irrigation	2+3+4+1=5	1+0	6	3+3+1=4	1+0	5	6	6	Investigate further and change soon	II
Plant protection	2+3+3+2=4	1+0	5	3+4+1=5	0+0	5	6	6	Investigate further and change soon	II
Harvesting	5+3+4+1=7	1+0	8	4+5+1=7	0+0	7	7	7	Investigate and change immediately	I
Handling Transportation	4+3+4+1=6	1+0	7	3+3+1=4	0+0	4	6	6	Investigate further and change soon	II

Pruning : The mean value of wrist and arm was 8 and muscles used forces was 1, thus giving the total score of 8. The mean score for neck, trunk and leg was worked out to be 9 with muscle force of 1, and total score (score B) of 7. This score of 7 indicates action category investigate and change in posture immediately in pruning.

Harvesting : The mean value of wrist and arm score was 8 and muscles used force was 1, thus giving the total score of 8. The mean score for neck, trunk and leg was worked out 7 with muscle force of 0, and total score (score B) of 7. This scores generate action level which implies that the investigate and immediately change in the posture used for harvesting.

Manuring : The mean value of wrist and arm score was 4 and muscle used force was 0, thus giving the total score of 4. The mean score for neck, trunk and leg was worked out 5 with muscle force of 0, total score (score B) of 5. This score of 5 implies that investigate further the posture used in manuring and change it soon.

Irrigation : The mean value of wrist and arm score was 5 and muscle used force was 1, thus giving the total score of 6. The mean score for neck, trunk and leg was worked out with muscle force of 1 resulting in total score (score B) of 6. This score of 6 indicates that further investigate the posture used in irrigation and change it soon.

Plant protection : The mean value of wrist and arm score was 4 and muscle used was 1, thus giving the total core of 5. The mean score for neck, trunk and leg was worked out with muscle force of 0 and total score (score B) of 6. This score generates action investigate further the posture used and change it soon.

Handling and transportation : The mean value of wrist and arm score was 6 and muscle used 1, thus giving the total score of 7. The mean score neck, trunk and leg with muscle force of 0 was total score (score B) of 6. This score generates action investigate further the posture used and change soon.

As per the RULA of activities in grape orchard posture used in land preparation, pruning and harvesting required investigation and should be changed immediately.

Ovako working posture analysis (OWAS) of activities in grapes cultivation

The ovako working posture analysis of workers in grapes cultivation was assessed and presented in Table 4.24.

Table 4.24: Ovako working posture analysis (OWAS) of activities in grapes cultivation

(n=15)

Activity	Activity score				OWAS score	OWAS action category	Rank
	Back	Arms	Legs	Load / use of force			
Land preparation	2	1	5	1	**3**	Corrective measures as soon as possible	II
Pruning	3	2	4	2	**4**	Corrective measures immediately	I
Manuring	4	1	7	2	**3**	Corrective measures as soon as possible	II
Irrigation	2	1	4	2	**3**	Corrective measures as soon as possible	II
Plant protection	1	3	2	1	**1**	No corrective measures	III
Harvesting	4	2	4	1	**4**	Corrective measures immediately	I
Handling and transportation	4	1	2	2	**2**	Corrective measures in the near future	II

Land preparation : The activity score of back (2), arm(1), legs (5) and load force used (1) resulted in OWAS score of 3 depicting that in land preparation corrective measure in posture was required as soon as possible.

Pruning : The activity score of back (3), arm (2), legs (4) and load force used (2) resulted in OWAS score of 4 depicting that pruning required corrective measures immediately in the posture used.

Manuring : The activity score of back (4), arm (1), legs (7) and load force used (2) resulted in OWAS score of 3 depicting that posture in manuring required corrective measures as soon as possible.

Irrigation : The activity score of back (2).arm (1), legs (4) and load force used (2) resulted in OWAS score of 3 depicting that irrigation required corrective measures of posture as soon as possible.

Plant protection : The activity score back (1), arm (3), legs (2) and load force used (1) resulted in OWAS score of 1 depicting that in plant protection required no corrective measures in posture used.

Harvesting : The activity score back (4), arm (2), legs (4) and load force used (1) resulted in OWAS score of (4) depicting that posture used in harvesting required corrective measures immediately.

Handling and transportation : The activity score back (4), arm(1), legs (2) and load force used(2) resulted in OWAS score of 2 depicting that corrective measures was needed in the posture used in near future.

OWAS analysis of activities of grape orchard depicted that posture used in pruning and harvesting was highly risky and corrective measures in posture was required immediately.

Posture load score of workers in grapes cultivation

The posture load of the workers in grapes cultivation was assessed and presented in table 4.25 and Fig 4.8.

Land preparation : Posture load factor during land preparation for the removing of stalks and stubbles and unwanted plants was 30.6 and ploughing by country plough was 33.5, whereas overall posture load was 32.05.

Pruning : Posture load factor for pruning included cutting of undesirable vines and it was 47.8.

Manuring : Activities under manuring were transportation of manure with posture load factor 25, mixing of manure with posture load factor 29.6, spreading of manure with posture load factor 35 and mean posture load factor was 29.8.

Irrigation : Posture load factor during irrigation for preparation of irrigation of channels was 25.

Plant protection : Among plant protection covering with net got posture load factor of 25, spraying and topping, each got posture load factor of 27. Mean posture load factor in plant protection was 26.3.

Harvesting : The activities covered under harvesting were fruit picking with posture load factor of 37.7, gathering and heaping with posture load factor 18.7 packaging in polythene with posture load factor 30.2 and trimming with posture load factor 29.6. Mean posture load of harvesting was 29.05.

Handling and transportation : Posture load factor in handling and transportation includes loading of the product and posture load factor was 24.8.

The posture load factor was highest in pruning activity indicating that pruning was highly risk prone activity in grapes cultivation.

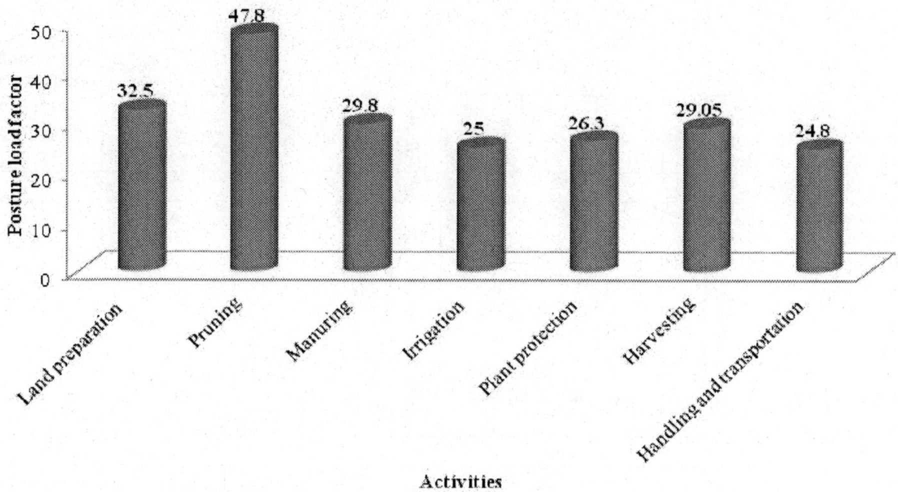

Fig. 4.8: Posture load score of workers in grapes cultivation

Repetitive Strain (exertion)- The repetitive strain in grape cultivation was assessed through:

- ART tool (assessment of repetitive task)Health and safety executive (2010)
- Strain Index (Moore, JS and Garg A, 1995)
- Repetitive Strain load score

Assessment of repetitive tasks (ART) of the upper limb in grapes cultivation activities

The various aspects under assessment of repetitive tasks (ART) of the upper limbs were A1 arm movement, A2 repetition, B force, C1 head neck posture, C2 back posture, C3 arm posture, C4 wrist posture. C5 hand finger, D1 breaks, D2 work pace, D3 other factor D4 duration and task score. The table 4.26 depicts the ART in various activities of grapes cultivation.

Land preparation : Assessment of repetitive tasks of upper limb score for left hand was 25 and right hand was 25, resulting in overall score of 50.The exposure score for both hands generates that further investigation was required in the task of land preparation.

Table 4 25: Posture load score of workers in grapes cultivation

(n=15)

Farm Activity	Nature of posture	Body part involved													Posture load factor (Total Score)	Mean posture load score
		Head	Neck	Shoulder	Upper arms	Lower arms	Fingers	Hands	Waist	Hips	Upper legs	Lower legs	Upper back	Lower back		
Land preparation																
Removing of stalks and stubbles and unwanted plants	4	-	1.8	1.2	1.4	3.6	2.6	3.6	1.2	1.2	1.8	2.7	2.4	2.8	**30.6**	**32.05**
Ploughing by country plough	2	1	1.8	4.6	4.1	3.8	1.7	2.8	1	1	2	1.6	2.6	3.4	**33.5**	
Pruning																
Cutting of undesirable vines	2.0	1.9	2.0	3	3.8	5	4.8	5	2	1	4	5	4	4	**47.8**	**47.8**
Manuring																
Transportaions of manure	1	1	1.4	1.8	1.5	1.4	1.4	1.9	1	1	2.2	2.6	3.0	3.4	**25**	**29.8**
Mixing of manure	6	1	1	2	2.6	2	2	3	1	1	2	1	2	3	**29.6**	
Spreading of manure	1	1	2	3	4	3	3	4	1	1	3	2	3	4	**35**	
Irrigation																
Preparation of irrigation	1	1	1.4	1.8	1.5	1.4	1.4	1.9	1	1	2.2	2.6	3.0	3.4	**25**	**25**
Plant protection																
Spraying	1	1	1	1	3	2	2	2	1	1	4	3	3	2	**27**	
topping	1	1	1	1	3	2	2	1	1	4	3	3	3	2	**27**	
Harvesting																
Fruit picking	2	1	1.5	2.7	4.0	3.1	3.3	3.2	1	1	3.3	3.7	3.6	3.9	**37.7**	**29.05**
Gathering and heaping	4	1	1.0	1	1	1	1	1	1	1	1	2.6	1	1	**18.7**	
Packaging in polythene	4	1	1.7	2	2.4	3.2	3	3	1	1	1	1	2.4	3.3	**30.2**	
Trimming	6	1	1	2	2.6	2	2	3	1	1	2	1	2	3	**29.6**	
Handling and transportation																
Loading of the product	1	1	1.5	3.6	3.3	2.6	1.4	1.6	1	1	1.4	1.2	1.6	2.2	**24.8**	**24.8**

Nature of posture: Standing-1, Bending-2, Twisted trunk-3, Sitting-4, Squating-5, Kneeling-6

Pruning : For pruning the assessment of repetitive tasks score for left hand was 28 and right hand was 30, overall score was 58. The exposure score for both hands indicates that further investigation was required urgently in the pruning task.

Manuring : During manuring assessment of repetitive tasks of upper limb ART score for left hand was 4.5 and that of right hand was 11.25, resulting in overall score of 15.7. The exposure score for both hands depicted that consideration of individual circumstance (Both hands) was required in manuring.

Irrigation : The ART score for left hand was 3.75 and right hand was 6.75 in irrigation and overall score was 10.5. This exposure score highlighted for consideration of individual circumstance (both hands) in irrigation.

Plant protection : In plant protection activity, the ART score for left hand was 2.25, right hand was 2.25, resulting in overall score of 4.5. The exposure score generated depicted that consideration of individual circumstance (both hands) was required in the plant protection.

Harvesting : The ART tool score in harvesting for left hand was 24 and right hand was 26 and overall score was 50, the exposure score generated indicated that further investigation was required urgently for harvesting activity.

Handling and transportation : The ART score for left hand was 19 and right hand 20 and overall score was 39, which depicted that further investigation was required urgently in handling and transportation activity.

As per the assessment of repetitive tasks (ART), in grapes cultivation pruning was risky activity and further investigation was required immediately.

Table 4.26: ART (tools) assessment of repetitive tasks of the upper limb in grapes cultivation activities

(n=15)

	A1 (Arm movement)		A2 (Repetition)		B (Force)	C1 (Head/ Neck posture)	C2 (Back posture)	C3 (Arm posture)		C4 (Wrist posture)		C5 (Hand / finger grip)		D1 (Breaks)	D2 (Workspace)	D3 (other factor)		D4 (Duration)	(Task score)	Exposure score	Ranking
	L	R	L	R				L	R	L	R	L	R			L	R		A1+A2+B+C1+C2 +C3+C4+C5+D1+ D2+D3		
Land preparation	3	3	3	3	R12	1	1	2	2	1	1	0	0	0	1	1	1		Left hand= 3+3+r12+1+1+2+1 +0+1+1=25 Right hand= 3+3+R12+1+1+2+1 +0+0+1+1= 25×1=25 **Total= 50**	Further investigation required (Both hands)	II
Pruning	3	3	3	3	R_{12}	2	2	2	2	1	2	0	0	0	2	1	2	×1	Left hand= 3+3+R12+2+2+2+1 +0+0+2+1= 28×1=28 Right hand= 3+3+R12+2+2+2+2 +0+0+2+2= 30×1=30 **Total =58**	Further investigation required urgently	I
Manuring	0	3	0	3	A4	0	0	0	0	0	1	0	1	2	0	0	1	×0.75	Left hand= 0+0+A4+0+0+0+0+ 0+2+0)= 6×0.75=4.5 Right hand= 3+3+A4+0+0+0+1+ 1+2+0+1= 15×0.75=11.25 **Total = 15.7**	Consider individual circumstance (Both hands)	IV
Irrigation	0	0	0	3	A2	0	1	0	0	0	0	0	1	2	0	0	0	×0.75	Left hand= 0+0+A2+0+1+0+0+ 0+2+0+0= 5×0.75=3.75 Right hand= 0+3+A2+0+1+0+0+ 1+2+0+0= 9×0.75=6.75 **Total = 10.5**	Consider individual circumstance (Both hands)	V
Plant protection	0	0	0	0	A1	0	0	0	0	0	0	0	0	2	0	0	0	×0.75	Left hand= 0+0+A1+0+0+0+0+ 0+2+0+0= 3×0.75=2.25 Right hand= 0+0+A1+0+0+0+0+ 0+2+0+0= 3×0.75=2.25 **Total = 4.5**	Consider individual circumstance (Both hands)	VI
Harvesting	3	3	6	6	A4	2	2	2	2	1	1	1	2	0	2	1	2	×1	Left hand= 3+6+A4+2+2+1+1+ 0+2+1= 24×1=24 Right hand= 3+6+A4+2+2+2+1+ 2+0+2+2= 26×1=26 **Total = 50**	Further investigation required urgently	II
Transportation	0	0	6	6	A4	1	1	2	2	1	1	0	1	2	1	1	1	×1	Left hand= 0+6+A4+1+1+2+1+ 0+2+1+1= 19×1=19 Right hand= 0+6+A4+1+1+2+1+ 1+2+1+1= 20×1=20 **Total = 39**	Further investigation required urgently	III

Strain Index analysis of activities in grapes cultivation

The various aspects under strain index analysis were intensity of exertion, duration of exertion (% of cycle), efforts per minute, hand wrist/posture, speed of work, duration of task(per day)and strain index and presented in Table 4.27.

Land preparation : The activity score of intensity of exertion (6.0), duration of exertion (1.5), efforts per minute(1.0), hand wrist posture (1.5), speed of work (1.5), duration of task (per day) (1.0) resulted in strain index of 6.8, which indicate that in land preparation there was some risk/strain.

Pruning : In pruning activity, score of intensity of exertion (13.0), duration of exertion (3.0), efforts per minute(2.0), hand wrist posture (2.0), speed of work (1.5), duration of task (per day) (0.8) which gave strain index of 175.5 depicting that pruning was hazardous activity.

Manuring : For manuring the activity score of intensity of exertion (3.0), duration of exertion (1.0), efforts per minute(1.0), hand wrist posture (1.5), speed of work (1.0), duration of task (per day) (0.8) resulted in strain index of 3.4, depicting that in manuring the level of strain was uncertain.

Irrigation : The activity score of intensity of exertion (3.0), duration of exertion (1.0), efforts per minute(1.0), hand wrist posture (1.5), speed of work (1.0), duration of task (per day) (0.8) in irrigation resulted in strain index of 3.4 depicting that strain in irrigation was uncertain.

Plant protection : In plant protection, the activity score of intensity of exertion (3.0), duration of exertion (1.0), efforts per minute(1.0), hand wrist posture (1.5), speed of work (1.0), duration of task (per day) (0.8) gave strain index of 3.4, depicting that strain in plant protection was uncertain.

Harvesting : For harvesting the activity score of intensity of exertion (6.0), duration of exertion (2.0), efforts per minute(2.0), hand wrist posture (2.0), speed of work (1.5), duration of task (per day) (0.8) gave strain index of 54.0, which highlighted that harvesting was hazardous activity.

Handling and transportation : The activity score of intensity of exertion (6.0), duration of exertion (2.0), efforts per minute(2.0), hand wrist posture (2.0), speed of work (1.5), duration of task (per day) (0.8) resulted in strain index of 54.0,depicting that strain in handling and transportation was hazardous activity.

As per strain index of the activities of grape cultivation, pruning harvesting and handling transportation were hazardous activities but among these pruning score highest score.

Table 4.27: Strain Index analysis of activities in grapes cultivation

(n=15)

Grapes orchards activity	Intensity of Exertion	Duration of Exertion (% of cycle)	Efforts per minute	Hand wrist / posture	Speed of work	Duration of task (Per day)	Strain Index	Interpretation	Ranking
Land preparation	6.0	1.5	1.0	1.5	1.5	1.0	6.8	Some risk	III
pruning	13.0	3.0	2.0	2.0	1.5	0.8	175.5	Hazardous	I
Manu ring	3.0	1.0	1.0	1.5	1.0	0.8	3.4	Uncertain	IV
Irrigation	3.0	1.0	1.0	1.5	1.0	0.8	3.4	Uncertain	IV
Plant protection	3.0	1.0	1.0	1.5	1.0	0.8	3.4	uncertain	IV
Harvesting	6.0	2.0	2.0	2.0	1.5	0.8	54.0	Hazardous	II
Handling and transportation	6.0	2.0	1.5	2.0	1.5	1.0	54.0	Hazardous	II

Repetitive strain load scores of workers in grapes cultivation

Table 4.28 and Fig 4.9 presents the repetitive strain load factor in grapes cultivation.

Land preparation : Repetitive strain load score during various activities land preparation was 3.8 for the removing of stalks and stubbles and unwanted plants and 4.6 for ploughing by country plough. Mean repetitive strain score for land preparation was 4.2.

Pruning : Repetitive strain load factor during pruning for cutting of undersirable vines was 6.

Manuring : During manuring, transportation of manure got repetitive strain load score was 4.6 and mixing of manure got 3, spreading of manure got 4.8. The mean repetitive strain load factor in manuring was 4.1.

Irrigation : Repetitive strain load score during irrigation for preparation of irrigation channels was 4.4.

Plant protection : Activities among plant protection includes covering with net spraying and topping, which got repetitive strain load score of 4 each. Mean repetitive strain load score of plant protection was 4.

Harvesting : Fruit picking got repetitive strain load score of 5.2, gathering and heaping got 4 packaging in polythene got 4.4 and trimming got repetitive strain load factor of 4 and mean repetitive strain load factor of harvesting was 4.4.

Handling and transportation : Repetitive strain load score in handling and transportation for loading of the product was 6.

The most repetitive activity in grape cultivation was pruning, as it got highest repetitive strain load factor, hence risky in terms of repetitive nature of activity.

Table 4.28: Repetitive strain load score of workers in grapes cultivation

(n=15)

Farm activity	Nature	Repetitive Strain Rating[3]	Repetitive strain load score	Mean Repetitive strain load score
Land preparation				
Removing of stalks and stubbles and unwanted plants	1	2.8	3.8	4.2
Ploughing	1	3.6	4.6	
Pruning				
Cutting of undesirable vines	1	5	6	6
Manuring				
Transporation of manure	2	2.6	4.6	4.1
Mixing of manure	1	2	3	
Spreading of manure	2	2.8	4.8	
Irrigation				
Preparing channel for irrigation	1	3.4	4.4	4.4
Plant protection				
Covering with net	1	3	4	4
Spraying	1	3	4	
Topping	1	3	4	
Harvesting				
Fruit picking	1	3.2	5.2	4.4
Gathering and heaping	1	3	4	
Packaging in polythene	1	3.4	4.4	
Trimming	1	3	4	
Handling and transportaion				
Loading of the product	2	4	6	6

Repetitive strain nature- Continuous-1, Cyclic-2

Repetitive Strain rating- Very exhausted-5, Exhausted-4, Moderately exhausted-3, Mildly exhausted-2, Comfortable-1

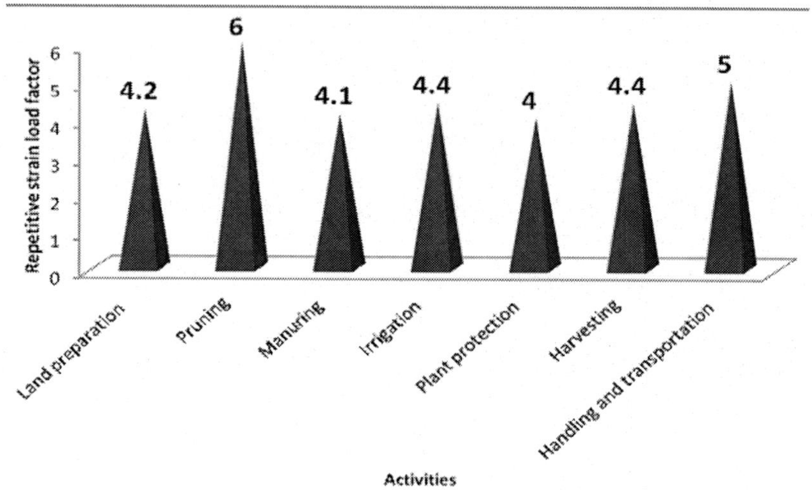

Fig. 4.9: Repetitive strain load of score workers in grapes cultivation

Total Repetitive strain (exertion) of workers in grape cultivation

Total repetitive strain score was assessed in terms of ART tool, strain index, repetitive strain load and presented in table 4.29. The repetitive strain load was maximum in pruning with total repetitive strain load (239.5), followed by harvesting with total repetitive strain load (108.4). The total repetitive strain load in decreasing order in other activities was handling and transportation (99), land preparation (61), manuring (23.2), irrigation (18.3), plant protection (11.9).

On the basis of total repetitive strains score in grapes cultivation pruning was most risky activity followed by harvesting.

Table: 4.29 Total Repetitive strain (exertion) of workers in grapes cultivation

Activities	Repetitive strains			Total repetitive strain (exertion) score	Rank
	ART Score	Strain index score	Repetitive strain load score		
Land preparation	50	6.8	4.2	61	IV
Pruning	58	175.5	6	239.5	I
Manuring	15.7	3.4	4.1	23.2	V
Irrigation	10.5	3.4	4.4	18.3	VI
Plant protection	4.5	3.4	4	11.9	VII
Harvesting	50	54.0	4.4	108.4	II
Handling and transportation	39	54.0	6	99	III

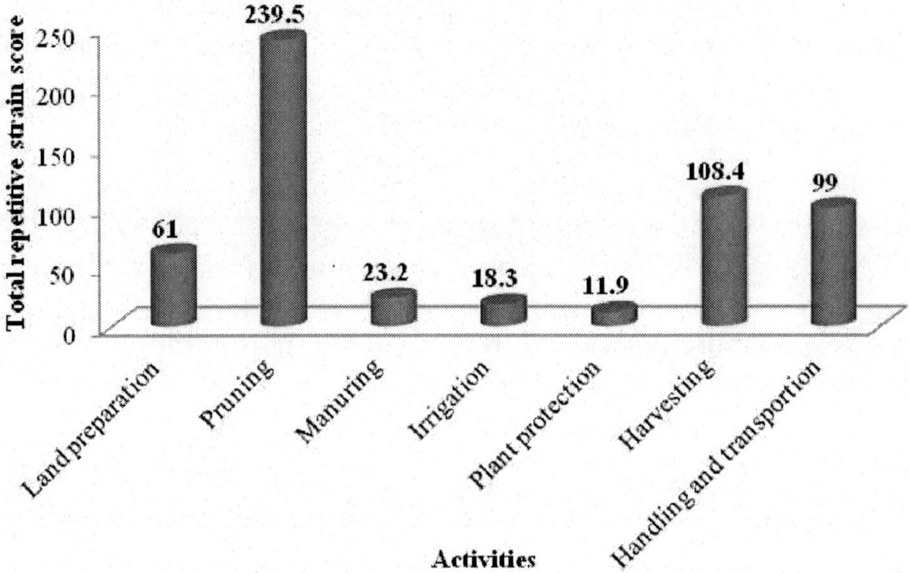

Fig. 4.10: Total Repetitive strain (exertions) of workers in grapes cultivation

Musculoskeletal disorder / discomfort

Musculoskeletal discomfort in activities of grape cultivation was assessed through:

- Human body Map (Corlett & Bishop. 1976)
- VAD Scale (ODS) (Corlett and Bishop, 1976)
- NORDIC Scale, (Kuorinka *et al.*, 1987)

Mean score of Intensity of musculo-skeletal pain in workers of grapes orchards as per body map

Intensity of Musculoskeletal pain of workers in grape cultivation as per body map has was recorded after performing the grapes cultivation activities on a five-point scale as per human body map as indicating very severe pain, severe pain, moderate pain, mild pain and very mild pain been presented in Table 4.30.

Land preparation : In land preparation activity, intensity of musculo-skeletal discomfort was highest in upper back and lower back with mean score 35.2 (rank II)

Pruning : For pruning maximum Intensity of musculo-skeletal discomfort in wrist and neck with mean score 34.4 (rank IV).

Manuring: During manuring, majority of the workers were suffering from lower back, wrist, hand with mean score 33.7 (rank VI)

Irrigation : Most of the workers were facing Intensity of pain in upper arm, buttock, wrist with mean score 4.2 (rank V)

Plant protection : For plant protection,maximum intensity of musculo-skeletal discomfort in lower back, lower arm,ankle/feet with mean score 34.4 (rank IV)

Harvesting : In harvesting,maximum workers were suffering from intensity of musculo-skeletal in hand, thigh, neck with mean score 35.7 (rank I)

Handling and transportation : In handling and transportation, intensity of musculo-skeletal discomfort was highest in shoulder, lower back, wrist with mean score 34.6 (rank III)

Table 4.30: Mean score of intensity of musculo-skeletal pain in workers of grapes orchards as per body map

(n=15)

Activities	Neck	Shoulder	Upper back	Upper arm	Lower back	Lower arm	Buttock	Wrist	Hand	Thigh	Feet	Total	Rank
Land preparation	3.2	3.2	3.6	3.2	3.5	3	2.8	3.6	3.5	2.8	2.7	35.2	II
Pruning	3.5	3.1	3.2	3.2	3.4	3.2	2.7	3.6	3.3	2.6	2.6	34.4	IV
Manuring	2.8	3.1	3.1	3.0	3.3	3.0	2.7	3.3	3.4	2.8	3.2	33.7	VI
Irrigation	3.2	2.9	3.2	3.3	3.0	2.7	3.3	3.4	2.8	3.2	3.2	34.2	V
Plant protection	2.9	2.6	2.6	2.8	3.2	3.2	2.9	2.6	2.6	2.6	3.2	34.4	IV
Harvesting	3.5	3.1	3.2	3.2	3.4	3.2	2.7	2.9	3.8	3.6	3.1	35.7	I
Handling and transportation	3.2	3.6	3	2.8	3.5	3.1	2.8	3.4	2.7	3.3	3.1	34.6	III

Visual Analogue Discomfort (VAD) of workers in various activities of grapes cultivation

Table 4.31 reflects the visual analogue discomfort scale in different activities of grapes production system. On the basis of observation, land preparation pruning and harvesting activity got VAD score of 8.0, hence these were found to be very vigorous activity. However, handling and transportation got VAD score of 5.0, hence depicting them moderate activity. Data further showed that manuring, irrigation and plant protection had VAD score of 3.0 indicating them as mild activities.

Land preparation, pruning and harvesting activities were the considered as the vigourous activities by the workers in grapes cultivation hence were risky as per VAD score.

Table 4.31: Visual Analogue Discomfort (VAD) of workers in different activities in grapes cultivation

(n=15)

Activity	Visual analogue discomfort (VAD) Score	Action	Ranking
Land preparation	8.0	**Vigorous activity**	1
Pruning	8.0	**Vigorous activity**	1
Manuring	3.0	Mild activity	3
Irrigation	3.0	Mild activity	3
Plant protection	3.0	Mild activity	3
Harvesting	8.0	**Vigorous activity**	1
Handling and transportation	5.0	Moderate activity	2

Nordic analysis of prevalence of musculo-skeletal problems in workers in grapes cultivation

Nordic analysis of prevalence of musculo-skeletal problems during 12 month of workers in grapes cultivation has been presented in Table 4.32.

Land preparation : In land preparation activity, prevalence of musculo-skeletal problems was highest in neck, wrist, lower back, one or both ankle/feet with mean score 12.2 (rank III).

Pruning : For pruning, maximum prevalence of musculo-skeletal problems in elbows, one or both ankle feet with mean score 12.7 rank (II).

Manuring : During manuring, majority of the workers were suffering from shoulder, elbows, wrist with mean score (11.4) rank IV

Irrigation : Most of the workers were facing prevalence of musculo-skeletal problems in shoulder, elbows with mean score 11.1 rank V.

Plant protection : For plant protection, maximum prevalence of musculo-skeletal problems in neck, elbows, lower back, one or both ankle/feet with mean score 12.7 rank II.

Harvesting : In Harvesting, maximum workers were suffering from prevalence of musculoskeletal problems in elbows, one or both ankle /feet with mean score 12.7 (rank II)

Handling and transportation : In handling and transportation prevalence of MSDs problem was higher in elbows and upper back with mean score 13.3 (rank I).

Table 4.32: Nordic analysis of prevalence of musculoskeletal problem of workers during 12 months in grapes cultivation

(n=15)

Activities	Mean score								
	Neck	Shoulder	Elbows	Wrist	Upper back	Lower back	One or both/ ankle/feet	Total	RANK
Land preparation	1.8	2	1.4	1.9	1.4	1.8	1.9	12.2	III
Pruning	2	2	1.8	2	1.5	1.5	1.9	12.7	II
Manuring	2	1.8	1.8	1.8	1.5	1.5	1	11.4	IV
Irrigation	2	1.8	1.8	1.5	1.5	1.5	1	11.1	V
Plant protection	1.9	1.7	1.9	2	1.3	1.9	1.8	12.7	II
Harvesting	2	2	1.8	2	1.5	1.5	1.9	12.7	II
Handling and transportation	2	2	1.8	2	1.8	1.7	2	13.3	I

Musculoskeletal discomforts of workers in grapes cultivation

The Overall musculo-skeletal discomfort was assessed on the basis of Human body map, VAD scale, Nordic scale presented in table 4.33. The musculoskeletal discomfort was maximum for harvesting with total score (56.4), followed by land preparation (55.4). The total score of other activities was pruning (55.1), handling and transportation (52.9), and plant protection (50.1). The least of total musculo-skeletal discomfort was for manuring (48.1) and irrigation (48.3).

Hence, musculo-skeletal discomfort in the grapes cultivation was highest for harvesting activity had highest MSD.

Table: 4.33 Musculo-skeletal discomforts of workers in grapes cultivation

Activities	Musculoskeletal discomforts			Total	Rank
	Human body Map Score	VAD Scale Score	Nordic scale Score		
Land preparation	35.2	8.0	12.2	55.4	II
Pruning	34.4	8.0	12.7	55.1	III
Manuring	33.7	3.0	11.4	48.1	VII
Irrigation	34.2	3.0	11.1	48.3	VI
Plant protection	34.4	3.0	12.7	50.1	V
Harvesting	35.7	8.0	12.7	56.4	I
Handling and transportation	34.6	5.0	13.3	52.9	IV

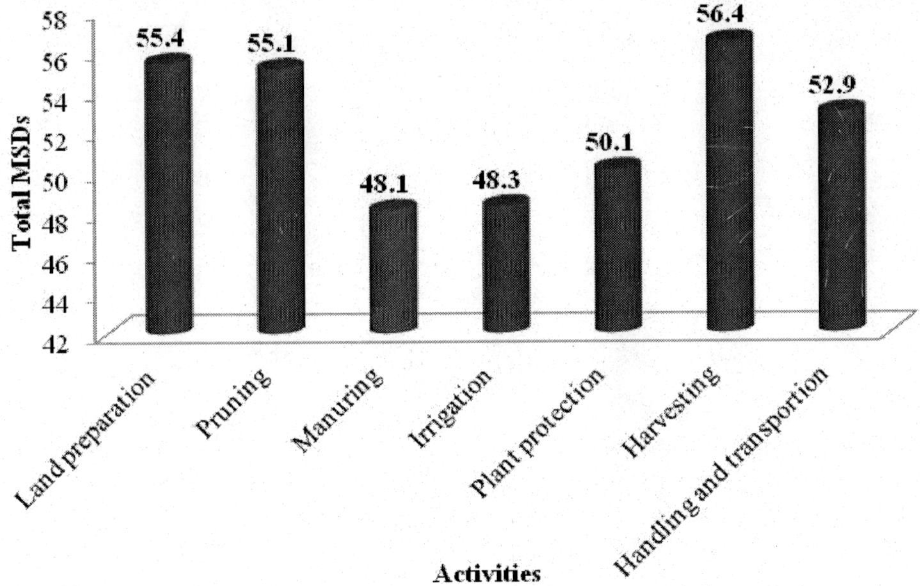

Fig. 4.11: Musculo-skeletal discomforts of workers in grapes cultivation

Occupational risks of workers in grapes cultivation terms of biomechanical stress

The occupational risks in grapes cultivation in terms of biomechanical stress was assessed on the basis of grip fatigue, RULA, OWAS, posture load score, repetitive strain exertion and musculo-skeletal discomfort and presented in Table 4.34. The occupational risks in terms of biomechanical stress was highest in pruning followed by harvesting. Further table 4.34 revealed that rank secured by other activities in descending order was handling and transportation (rank III), land preparation (rank IV), manuring (rank V), and irrigation (rank VI). The plant protection activity was least risky in terms of biomechanical stress.

Table 4.34: Occupational risks of workers in grapes cultivation terms of biomechanical stress

Activities	Biomechanical stress						Total of biomechanical stress	Rank
	Grip fatigue	RULA Score	OWAS Score	Posture load score	Repetitive exertion score	Musculoskeletal discomfort score		
Land preparation	29.9	7	3	32.0	61	55.4	188.3	IV
Pruning	30.9	7	4	47.8	239.5	55.1	384.3	I
Manuring	20.2	5	3	29.8	23.2	48.1	129.3	V
Irrigation	19.8	6	3	25	18.3	48.3	120.4	VI
Plant protection	11.2	6	1	26.3	11.9	50.1	106.5	VII
Harvesting	15	7	4	29.05	108.4	56.4	219.8	II
Handling and transportation	14.3	6	2	24.8	99	52.9	199	III

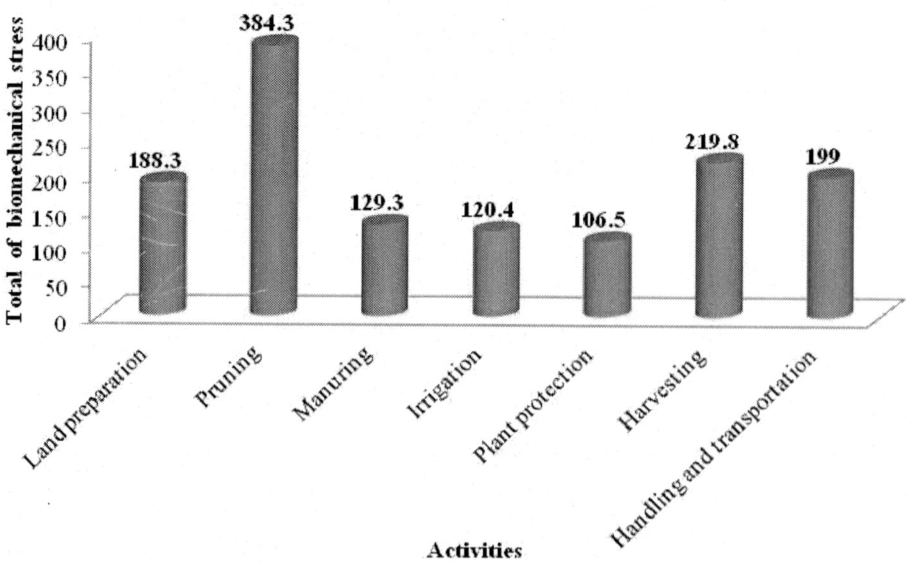

Fig. 4.12: Occupational risks of workers in grapes cultivation in terms of biomechanical stress

Occupational risks of workers in grapes cultivation in terms of Physiological Load

Occupational risk of workers in terms of physiological load in grapes cultivation was assessed through:

- Heart rate
- Oxygen consumption (l/min)
- Energy expenditure(KJ/min)
- Perceived Exertion
- Physiological load factor

Physiological load in grapes cultivation has been presented in Table 4.35.

Heart rate : Heart rate measurement during any activity is an indicator of cardiac stress due to physical workload. Table 4.35 revealed that maximum heart rate of the workers was during land preparation activity, which was 154.5 bpm and during pruning it was 119.2 bpm. Heart rate for harvesting, irrigation and manuring was 112 bpm, 107.1 bpm and 109.4bpm, respectively.

Energy expenditure KJ/min : Table 4.35 revealed that maximum energy expenditure of the workers was in land preparation which was 158.8 KJ/min and in pruning, it was 10.2 KJ/min. Energy expenditure in harvesting, manuring and irrigation was 9.13 KJ/min, 8.67KJ/min 8.30 KJ/min, respectively.

Oxygen expenditure rate ($l.min^{-1}$) : Maximum oxygen expenditure rate ($l.min^{-1}$) of the grapes orchards workers in land preparation activity was 0.97 l min^{-1} and in pruning, it was 0.62 $lmin^{-1}$. Oxygen consumption rate for harvesting, irrigation and manuring was 0.511 min^{-1}, 0.47 $lmin^{-1}$, 0.43 $lmin^{-1}$.

Table 4.35: Physiological parameters of workers before and after performing grapes cultivation activities

(n=15)

Activities	Parameters		
	Heart rate (bpm)	Energy expenditure (KJ/min)	Oxygen consumption rate ($1.min^{-1}$)
Land preparation	154.5±11.3	15.8±8.97	0.97±11.3
Manuring	109.4±11.3	8.67±8:97	0.43±11.3
Irrigation	107.1±11.3	8.30±8.97	0.47±10.6
Pruning	119.2±11.3	10.2±8.97	0.62±10.6
Harvesting	112±11.3	9.13±8.97	0.51±10.6

Association of age, frequency of rest period and BMI between physiological parameters

Data in Table 4.36 represents the association of age, frequency of rest period and body mass Index between physiological parameter of different grapes cultivation activities.

Heart rate

There was significant association of heart rate with age in land preparation, pruning and harvesting and non-significant association was found in manuring and irrigation activity of grapes cultivation. Similarly, in land preparation, pruning and harvesting a significant correlation was found in heart rate and frequency of rest period. However, in manuring and irrigation non-significant correlation existed in heart rate and frequency of rest period. A significant correlation existed between the heart rate and BMI, in all the activities of grapes cultivation viz., land preparation, pruning, manuring, irrigation and harvesting.

Energy expenditure

There was significant association of energy expenditure with age in land preparation, pruning and harvesting and non-significant association was found in manuring and irrigation activity of grape cultivation. Similarly, in land preparation, pruning and harvesting a significant correlation was found in energy expenditure and frequency of rest period non-significant correlation in manuring and irrigation. A significant correlation existed between the energy expenditure and BMI, in all the activities of grape cultivation viz, land preparation, pruning, manuring, irrigation and harvesting.

Oxygen consumption

There was significant association of oxygen consumption with age in land preparation, pruning and harvesting and non-significant association was found in manuring and irrigation activity of grape cultivation. Similarly, in land preparation, pruning and harvesting a significant correlation was found in oxygen consumption and frequency of rest period. However, in manuring and irrigation non-significant correlation was found a significant correlation existed between the oxygen consumption and BMI, in all the activities of grapes cultivation viz, land preparation, pruning, manuring, irrigation and harvesting.

Table 4.36: Association of age, frequency of rest period and BMI between Physiological parameter

Physiological load	Independent Variable		
	Age	Frequency of rest period	BMI
Heart Rate			
Land preparation	0.520*	0.515*	0.647**
Pruning	0.55*	0.050*	0.560*
Manuring	1.25^{NS}	0.348^{NS}	0.661**
Irrigation	0.288^{NS}	0.006^{NS}	0.549*
Harvesting	0.562*	0.641**	0.531*
Energy Expenditure			
Land preparation	0.514*	0.515*	0.641**
Pruning	0.525*	0.519*	0.562*
Manuring	0.125^{NS}	0.348^{NS}	0.661**
Irrigation	0.288^{NS}	0.006^{NS}	0.549**
Harvesting	0.549*	0.542*	0.515*
Oxygen Consumption			
Land preparation	0.559*	0.515*	0.548*
Pruning	0.66**	0.546*	0.524*
Manuring	1.25^{NS}	0.348^{NS}	0.661**
Irrigation	0.288^{NS}	0.006^{NS}	0.549*
Harvesting	0.514*	0.516*	0.539**

Rating of perceived exertion (RPE) of workers in different activities in grapes cultivation

Rating of perceived exertion is one of the most reliable tools for the subjective assessment of the exertion. Finding in table 4.37 highlighted the score of perceived exertion. Results showed that pruning activity got RPE score of 5 indicating that pruning was perceived as activity having very heavy exertion. RPE score of 4 was found for land preparation and harvesting activity, which means both that these activities were perceived as activity involving heavy excretion. Whereas irrigation, handling and transportation activity scored 3 which imply that these activities were considered as having moderate exertion. Manuring and plant protection were considered as activity having light exertion with RPE score of 2.

Table 4.37: Rating of perceived exertion of workers in different activities in grapes cultivation

(n=15)

Activity	Score	Action
Land preparation	4	Heavy exertion
Pruning	5	Very heavy exertion
Manuring	2	Light exertion
Irrigation	3	Moderate exertion
Plant protection	2	Light exertion
Harvesting	4	Heavy exertion
Handling and transportation	3	Moderate exertion

Physiological load of workers in grapes cultivation activities

The physiological load of workers in grapes cultivation was assessed and presented in Table 4.38.

Land preparation : Physiological load factor during various activities land preparation was 4 for the removing of stalks and stubbles and unwanted plants and 2.47 for ploughing by country plough. Mean physiological load factor for land preparation was 3.2

Pruning : Physiological load factor during pruning for cutting of undesirable vines was 5.

Table 4.38: Physiological load of workers in grapes cultivation activities

(n=15)

Farm activity	Physiological load rating[4]	Physiological load factor	Mean physiological load factor
Land preparation			
Removing of stalks and stubbles unwanted plants	4.00	4.00	3.2
Ploughing	2.47	2.47	
Pruning			
Cutting of undesirable vines	5.00	5.00	5
Manuring			
Transportation of manure	2.4	2.4	2.9
Mixing of manure	2.70	2.70	
Spreading of manure	3.83	3.83	
Irrigation			
Preparation of irrigation channels	4.00	4.00	4
Plant protection			
Covering with net	3.00	3.00	2.3
Spraying	2.00	2.00	
Topping	2.00	2.00	
Harvesting			
Fruit picking	4.27	4.27	3.5
Gathering and heaping	3.00	3.00	
Packaging in polythene	3.33	3.33	
Trimming	3.63	3.63	
Handling and transportation			
Loading of the product	3.67	3.67	3.67

Physiological work load rating : Very light-1, light-2, Moderately heavy- 3, Heavy-4, Very heavy-5

Manuring : During manuring transportation of manure got physiological load factor was 2.4 and mixing of manure got 2.70, spreading of manure got 3.83. The mean physiological load factor manuring was 2.9.

Irrigation : Physiological load factor during irrigation for preparation of irrigation channels was 4.

Plant protection : Activities among plant protection include spraying and topping, which got physiological load factor of 2 each. Mean physiological load factor of plant protection was 2.

Harvesting : Fruit picking got physiological load factor of 4.27, gathering and heaping got 3, packaging in polythene got 3.33 and trimming got physiological load factor of 3.63 and Mean physiological load factor of harvesting was 3.5.

Handling and transportation : Physiological load factor in handling and transportation for loading of the product was 3.67.

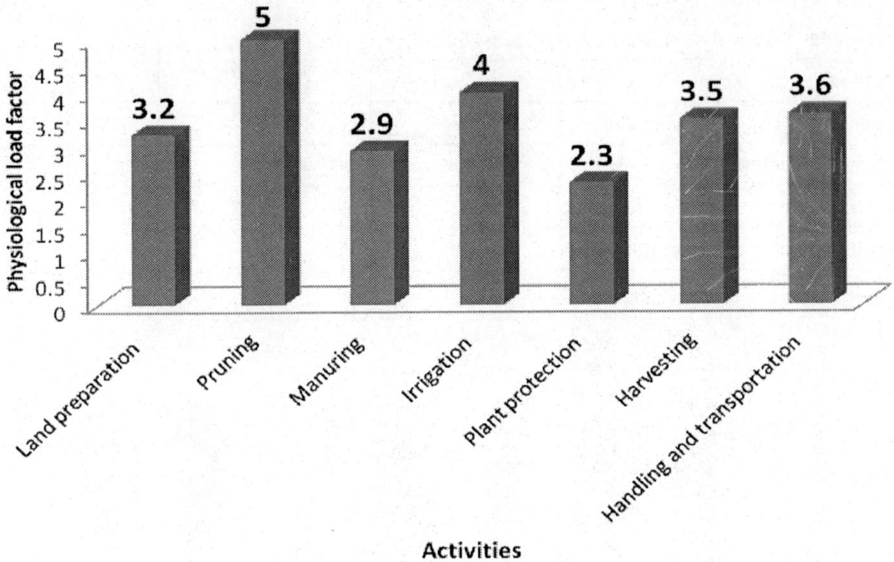

Fig. 4.13: Physiological load of workers in grapes cultivation activities

Occupational risks of workers in grapes cultivation terms of physiological load

Table 4.39 presents the occupational risks of workers in terms of physiological load. According to physiological load, pruning got I^{ist} rank and harvesting secured IInd rank. Land preparation obtained III rd rank followed by Irrigation which got IVth rank, further table 4.35 reveals that handling and transportation got (V) rank followed by manuring VI rank. Last rank was secured by plant protection (VII rank).

Table 4.39: Occupational risks of workers in terms of physiological load

Activities	Physiological load			Rank
	RPE	Physiological load factor	Total	
Land preparation	4	3.2	7.2	III
Pruning	5	5	10	I
Manuring	2	2.9	4.9	VI
Irrigation	3	4	7	IV
Plant protection	2	2.3	4.3	VII
Harvesting	4	3.5	7.5	II
Handling and transportation	3	3.67	6.67	V

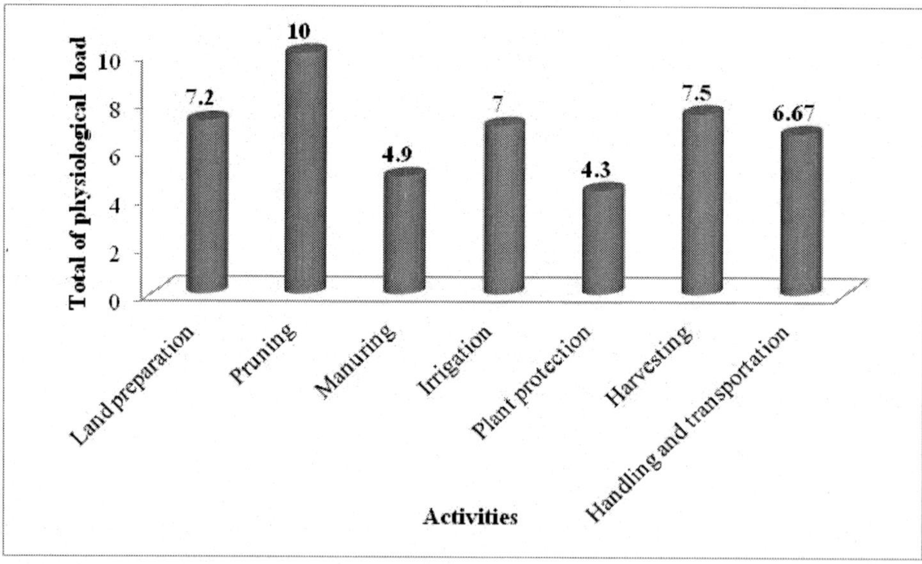

Fig. 4.14: Occupational risks of workers in terms of physiological load

IV Time load of workers in grape cultivation

Table 4.40 presents the time load factor of workers in grapes cultivation activities.

Land preparation : Time load factor during various activities of land preparation was 26.5, for removing of stalks and stubbles and unwanted plants and 14 for ploughing by country plough. Mean Time load factor for land preparation was 20.2.

Pruning : Time load factor during pruning for cutting of undersirable vines was 89.

Manuring : During manuring transportation of manure got time load factor was 11 and mixing of manure got 20, spreading of manure got 15. The Mean time load factor for manuring was 15.3.

Irrigation : Time load factor during irrigation for preparation of irrigation channels was 25.

Plant protection : Activities among plant protection includes spraying and topping which got time load factor of 32.25 and 58 respectively. Mean time load factor of plant protection was 34.7.

Harvesting : Fruit picking got time load factor of 112, gathering and heaping got 65.6, packaging in polythene got 62.6 and trimming got time load factor of 111 and mean time load factor of harvesting was 87.8.

Handling and transportation : Time load factor in handling and transportation for loading of the product was 58.6.

Pruning was most time consuming activity followed by harvesting.

Table 4.40: Time load of workers in grapes cultivation

(n=15)

Farm Activity	Duration / Time						Mean Time load factor
	Hours/ day	No. of days	No. of man days/ season[5]	No. of labour employed	Work load as per time[6]	Time Load Factor (Total score)	
Land preparation							
Removing of stalks and stubbles unwanted plants	8	7	3.5	5	3	**26.5**	**20.2**
Ploughing	8	1	1	2	2	**14**	
Pruning							
Cutting of undesirable vines	8	31	31	15	4	**89**	**89**
Manuring							
Transportation of manure	2	4	1	2	2	**11**	**15.3**
Mixing of manure	4	8	4	2	2	**20**	
Spreading of manure	1	8	1	2	3	**15**	
Irrigation							
Preparation of irrigation channels	8	4	4	5	4	**25**	**25**
Plant protection							
Covering with net	4	2	1	4	3	**14**	**34.7**
Spraying	5	10	6.25	7	4	**32.25**	
Topping	8	20	20	6	4	**58**	
Harvesting							
Fruit picking	8	45	45	10	4	**112**	**87.8**
Gathering and heaping	1	45	5.6	10	4	**65.6**	
Packaging in polythene	1	45	5.6	8	3	**62.6**	
Trimming	8	45	45	10	3	**111**	
Handling and transportation							
Loading of the product	1	45	5.6	4	3	**58.6**	**58.6**

No. of man days: 8hrs=1 man day

Work load as per time- Very high duration-5, High duration- 4, Moderate- 3, Less duration- 2, Very less duration-1

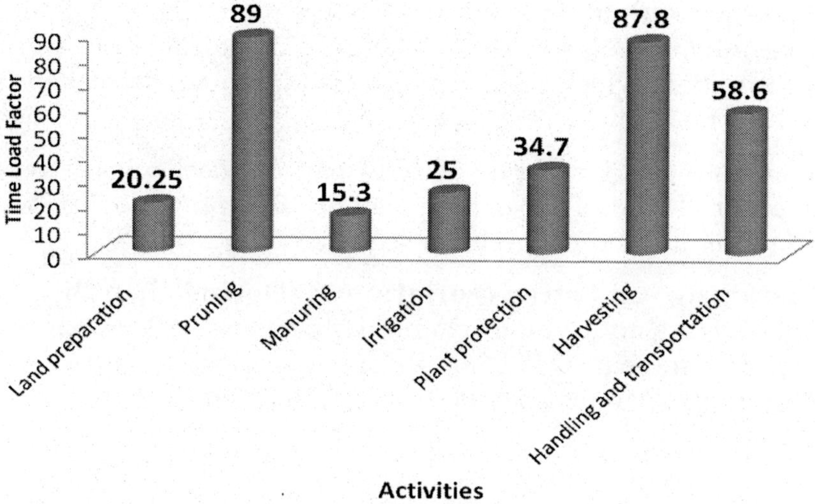

Fig. 4.15: Time load of workers in grapes cultivation

Overall Occupational risks of workers in grapes cultivation

The overall occupational risk of workers in grape cultivation was analyzed on the basis of physical load, biomechanical stress, physiological stress and time load and presented in 4.41 and Fig. 4.16.

Land preparation : In Land preparation, biomechanical stress score was 188.3, time load was 20.2, physical load was 100.4 and physiological load was 7.3 resulted in overall occupational risk of 304.7 with rank IV.

Pruning : In Pruning activity biomechanical stress score. Physical load, time load and, physiological load was 38.4.3, 178, 89 and 10 respectively which resulted in overall occupational risk of 640.2 and rank I.

Manuring: In Manuring activity biomechanical stress score was 129.3, physical load was 97.2, time load was 15.3 and physiological load was 4.9 which resulted in overall occupational risk of 256.6 and rank VI.

Irrigation : During Irrigation activity score of biomechanical stress physical load, time load and physiological load was 120, 81.5.25 and 7, respectively resulting in overall occupational risk of 205.9 with rank V.

Plant protection : Biomechanical stress score was 106.5, physical load was 61.5, time load was 34.7 and physiological load was 4.3 in plant protection, which resulted in overall occupational risk of 205.9 and rank VII.

Harvesting : Harvesting activity had Biomechanical stress of 219.5, physical load of 153.0, time load of 87.8 ,physiological load of 7.5, resulting in occupational risk score 467.8 and rank II.

Handling and transportation : During handling and transportation activity biomechanical stress was 199, physical load was 127.0, time load was 58.6, physiological load was 6.67, which resulted in occupational risk of 395.3 and rank III.

On the basis of overall occupational risk in grapes cultivation activities, pruning and harvesting were most risky activities. Hence, ergo solution developed for these activities.

Table 4.41: Overall Occupational risks of workers in grapes cultivation

Activities	Occupational risks				Overall occupational risks	Rank
	Physical load	Biomechanical stress	Physiological load	Time load		
Land preparation	100.4	188.3	7.2.0	20.2	316.1	IV
pruning	178	384.3	10.00	89	661.3	I
Manuring	97.2	129.3	4.9	15.3	246.7	VI
Irrigation	81.5	120.4	7.00	25	233.9	V
Plant protection	61.5	106.5	4.3	34.7	207.0	VII
Harvesting	153.05	219.8	7.50	87.8	460.7	II
Handling and transportation	127.05	199	6.67	58.6	391.2	III

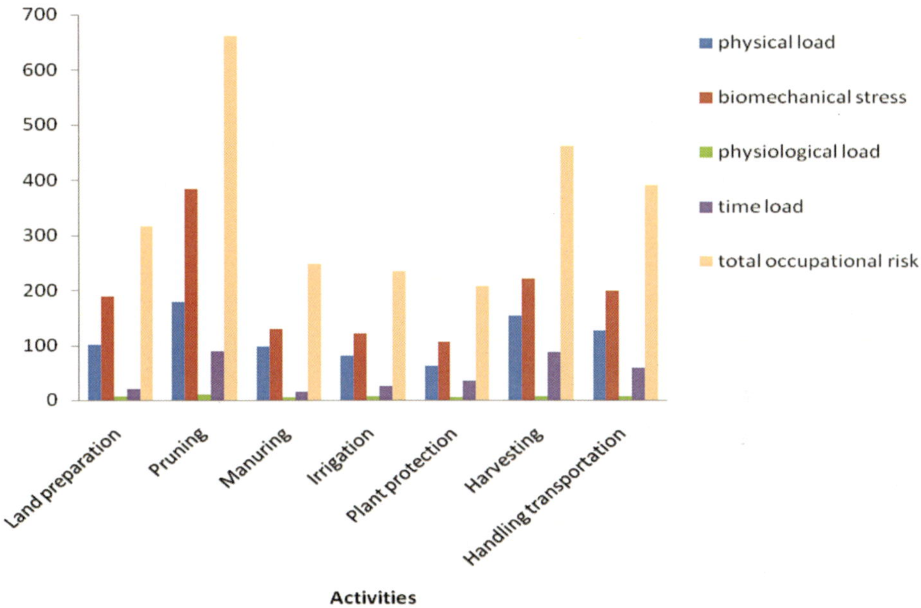

Fig. 4.16: Overall Occupational risks of workers in grapes cultivation

Phase III: Development of an ergo solution for highly risk prone activity in grapes cultivation

On the basis of results of Phase II pruning and harvesting of the grapes were identified as the highly risk prone activities. So, available technology used by them was modified and ergo solution was developed for these two activities.

Available technology i.e scateurs was modified by adding the materials on grip / handle. And an ergo stool for pruning and harvesting activity was developed and was named as ergo stool for pruning and harvesting. The acceptability and feasibility of ergo stool for pruning and harvesting was assessed.

Acceptability of improved grape pruning secateurs by orchard workers

The Acceptability of improved grape pruning secateurs was obtained from 10 grape orchard workers. Ranks were computed on the basis of mean score obtained in the following parameters and has been present in Table 4.42.

'Like the appearance, 'pruning can be done in proper way' and 'multipurpose in usage secured' 1st rank with (mean score 3.0). 'Increased coverage area under harvesting' and 'pruning got 2nd rank with (mean score 2.9). 'Comfortable size and shape of the grip while using' secured 3rd rank with (mean score 2.8), followed by the 'good replacement of the existing tool and technique'(mean score 2.7), 'able to maintained comfortable body posture in standing position while using modified scateurs' (mean score 2.6), 'minimum twisting of trunk while doing the activity' (mean score 2.5), 'However, restrained movements of hands' scored mean score of 2.2 and last rank was secured by 'coarse hard and unsuitable grip' (mean score 1).

Table 4.42: Acceptability of improved grape pruning secateurs by orchard worker

(n=10)

S. No.	Parameters	Mean score	Rank
1	Like the appearance	3.0	1
2	Comfortable size and shape of the grip	2.8	3
3	Easy to work as hand grips were convenient	3.0	1
4	Soft and suitable grip	1.0	8
5	Minimum twisting of trunk while doing the activity	2.5	6
6	Maintained comfortable body posture at standing position	2.6	5
7	Restrained movements of hands	2.2	7
8	Increased coverage area under harvesting and pruning	2.9	2
9	Good replacement to the existing tool and technique	2.7	4
10	Pruning can be done in proper way	3.0	1
11	Multipurpose in usage	3.0	1
12	Acceptability for use	3.5	1

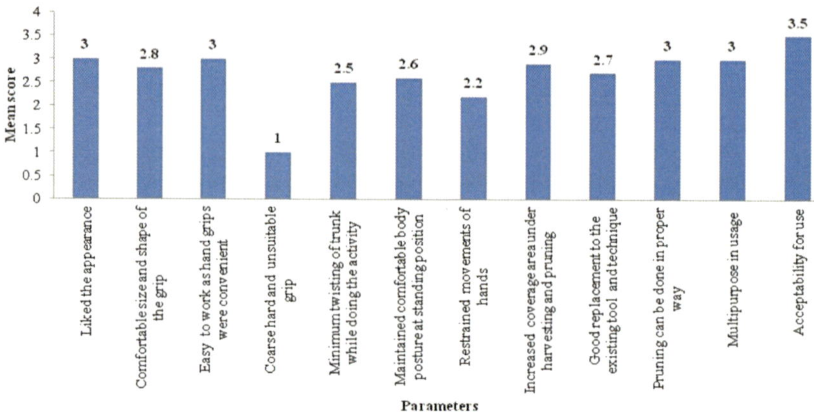

Fig. 4.17: Acceptability of improved grape pruning secateurs by orchard worker

Feasibility of improved grape pruning secateurs

Tables 4.43 depicts the feasibility of improved grape pruning secateurs in terms of profitability, physical compatibility, cultural compatibility, simplicity, traibility, grip fatigue, physical stress factor, work output, tool factor and acceptability.

Profitability : Table revealed that (80%) of the respondents considered improved grape pruning secateurs as most feasible, 10 per cent each considered it as feasible and somewhat feasible, technology for pruning of grapes in terms of profitability.

Physical compatibility : In terms of physical compatability, majority of the respondents (90%) considered improved grape pruning secateurs as most feasible and remaining 10 percent considered it as feasible technology.

Cultural compatibility : Table 4.43 revealed that regarding cultural compatability 40% of the respondents considered improved grape pruning secateurs as most feasible, 60 per cent respondents considered it as feasible technology.

Simplicity : Most of the respondents (80%) considered the improved grape pruning secater as most simple and rest of 20 per cent respondents considered it as simple technology for grape pruning.

Triability : Half of the respondents considered it as most feasible followed by 30 per cent who considered it as feasible and 20 per cent considered it as somewhat feasible technology for triability.

Grip fatigue : Table 4.43 further concluded that 80 per cent of the respondents considered grape pruning secateurs it as most feasible, 10 per cent each considered it as feasible and somewhat feasible, grape pruning technology in regards to grip fatigue.

Physical stress factor : Table 4.43 reveals that for physical stress factor 40% of the respondents considered grape pruning secateurs it as most feasible, half of the respondents considered it as feasible and 10 per cent respondents considered it somewhat feasible technology for grape pruning.

Work output : Majority of the respondents (80%) considered improved grape pruning secateur as most feasible and remaining 20 per cent respondents considered it as feasible technology for work output.

Tool factor : Most of the respondents 70% considered grape pruning secateurs as most feasible, and rest of (30%) respondents considered it feasible technology for pruning of grapes in terms of tool factor.

Acceptability : Cent percent of respondents considered grape pruning secateurs as most feasible, technology for grape pruning.

It is clear from table that, Overall feasibility of improved grape pruning secateurs was maximum (71.0%) as most feasible followed by feasible (24%) and somewhat feasible (5%), as grape pruning technology.

Table 4.43: Feasibility of improved grape pruning secateurs

(n=10)

S. No.	Feasibility	Most feasible	Feasible	Somewhat feasible
1.	Profitability	8(80)	1(10)	1(10)
2.	Physical compatibility	9(90)	1(10)	-
3.	Cultural compatibility	4(40)	6(60)	-
4.	Simplicity	8(80)	2(20)	-
5.	Triability	5(50)	3(30)	2(20)
6.	Grip fatigue	8(0)	1(10)	1(10)
7.	Physical stress factor	4(40)	5(50)	1(10)
8.	Work output	8(80)	2(20)	-
9.	Tool factor	7(70)	3(30)	-
10.	Acceptability	10(10)	-	-
	Overall n=100	**71(71.0)**	**24(24.0)**	**5(5.0)**

Feasibility index of improved grape pruning secateurs

Feasibility index was assessed in terms of profitability, physical compatibility, cultural compatibility, simplicity, triability, grip fatigue, physical stress factor, work output factor, tool factor and acceptability and presented in Table 4.44.

Feasibility Index : Tables 4.44 reveals that acceptability got first rank with feasibility index of 100%. Physical compatibility got 2nd rank with feasibility index of 98 per cent followed by simplicity and work output factor each with feasibility index of 96 per cent and 3rd rank, Grip fatigue and tool factor each got feasibility index 94 per cent with 4th rank. Cultural compatibility and profitability got 5th rank with feasibility index of 88 per cent. The last rank was for triability and physical stress factor with feasibility index of 86 per cent. Overall feasibility index of improved grape pruning sceateurs 93.2 per cent, which was on higher side indicate that it was accepatable to user.

Table 4.44: Feasibility index improved grape pruning secateurs

(n=10)

S.No.	Feasibility	Total	Feasibility index	Rank
1	Profitability	47	94	IV
2	Physical compatibility	49	98	II
3	Cultural compatibility	44	88	V
4	Simplicity	48	96	III
5	Triability	43	86	VI
6	Grip fatigue	47	94	IV
7	Physical stress factor	43	86	VI
8	Work output factor	48	96	III
9	Tool factor	47	94	IV
10	Acceptability	50	100	I
		466	**(93.2%)**	

Development of Ergo solution i.e Ergo stool for pruning and harvesting. Acceptability of Ergo stool for pruning and harvesting by grape orchard workers

The Acceptability of Ergo stool for pruning and harvesting was obtained from 10 grape orchard workers. Ranks were computed on the basis of mean score obtained and has been presented in Table 4.45.

Table unveils that "liked the appearance', acceptable for use got 1st rank with mean score 3. "Maintained comfortable body posture at standing position' and "shoulder pain can be avoided' got 2nd rank with mean score 2.9. 'Good replacement to the existing tool secured 3rd rank with mean score 2.7 'Convenient height of 'Ergo stool for harvesting and pruning "no pain and cramps in the upper region of body after performing the activity got rank 4 with mean score 2.6. Followed by convenient cleaning process with 5th rank and mean score 2.5 and "lesser discomfort in mid back region' secured 6th rank with mean score of 2.

Table 4.45: Acceptability of ergo stool for pruning and harvesting by the orchard workers

(n=10)

S. No.	Parameters	Mean score	Rank
1	Liked the appearance	3.0	I
2	Convenient cleaning process	2.5	V
3	Convenient height of ergo stool for pruning and harvesting	2.6	IV
4	Maintained comfortable body posture at standing position	2.9	II
5	Shoulder pain can be avoided	2.9	II
6	Lesser discomfort in mid back region	2.0	VI
7	No pain and cramps in the upper region of body after performing the activity	2.6	IV
8	Good replacement to the existing tool	2.7	III
9	Acceptability for use	3.0	I

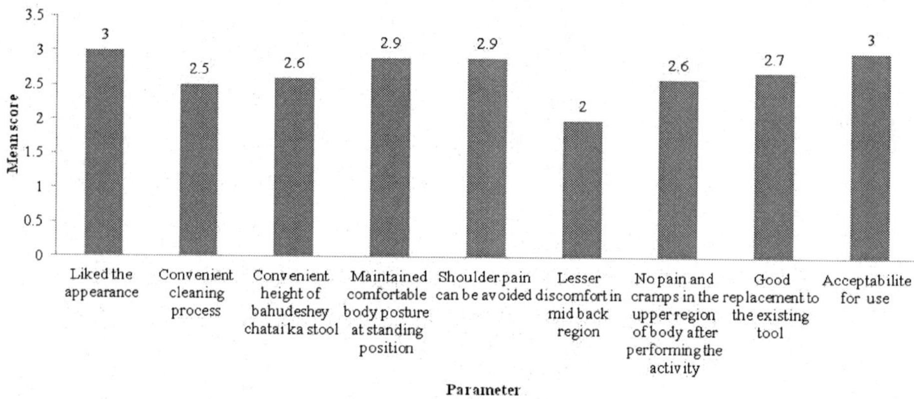

Fig. 4.18: Acceptability of ergo stool for pruning and harvesting by the orchard workers

Feasibility of Ergo stool for pruning and harvesting

Tables 4.46 depicts the feasibility of Ergo stool for pruning and harvesting in terms of profitability, physical compatibility, cultural compatibility, simplicity, triability, grip fatigue, physical stress factor, work output, tool factor and acceptability.

Profitabiltiy : Data presented in the Table 4.46 showed that cent per cent of the respondents considered a Ergo stool for pruning and harvesting profitable feasible technology.

Physical compatibility : Most of the respondents 30 % considered the Ergo stool for pruning and harvesting as feasible and while 70%

respondents considered it as somewhat feasible, technology for grape pruning, in terms of physical compatibility.

Cultural compatibility : For cultural compatibility 80 per cent of the respondents considered it as feasible, 20 per cent respondents considered it as somewhat feasible technology.

Simplicity : Cent per cent of the respondents considered Ergo stool for pruning and harvesting as most Simple feasible technology for pruning of grapes.

Triability : Cent per cent of the respondents considered *as* Ergo stool for pruning and harvesting feasible technology for traibility.

Grip fatigue : Table 4.46 further highlighted that for 60 per cent of the respondents Ergo stool for pruning and harvesting was most feasible and for 40.0 per cent respondents it as feasible technology for reducing the grip fatigue.

Physical stress factor : Table 4.46 revealed that cent percent of the respondents considered **Ergo stool for pruning and harvesting** as feasible technology for grape pruning for reducing physical stress.

Work output : Majority of the respondents (90%) considered Ergo stool for pruning and harvesting as feasible, 10 per cent respondents consider it as most feasible technology, to increase the work output.

Tool factor : All of the respondents considered it as most feasible technology in terms of tool factor.

Acceptability : Cent per cent respondents considered Ergo stool for pruning and harvesting *as* most feasible technology for the acceptability.

On the whole respondents 27 per cent considered Ergo stool for pruning and harvesting as most feasible 54 per cent respondents considered it as feasible, 19 per cent considered it as somewhat feasible, technology for pruning of grapes.

Table 4.46: Feasibility of Ergo stool for pruning and harvesting

S. No.	Feasibility	Most feasible	Feasible	Some what feasible
1	Profitability	-	10(100)	-
2	Physical compatibility	-	3(30)	7(70)
3	Cultural compatibility	-	8(80)	2(20)
4	Simplicity	10(100)	-	-
5	Triaibility	-	10(100)	-
6	Grip fatigue	6(60)	4(40)	-
7	Physical stress factor	-	10(100)	-
8	Work output	1(10)	9(90)	-
9	Tool factor	-	-	10(100)
10	Acceptability	10(100)	-	-
	Overall n=100	**27(27.0)**	**54(54.0)**	**19(19.0)**

Feasibility index of Ergo stool for pruning and harvesting

Feasibility index was assessed in terms of profitability, physical compatibility, cultural compatibility, simplicity, triability, grip fatigue, physical stress factor, work output factor, tool factor and acceptability and presented in Table 4.47.

It was clear from the table that simplicity and acceptability got first rank with feasibility index of 100%, physical stress factor and grip fatigue got 2nd rank with feasibility index 92 per cent followed by work output with feasibility index 82 per cent and 3rd rank. Profitability and triability secured feasibility index of 80 per cent and 4th rank. Culture compatibility got 5th rank with feasibility index of 76 per cent and tool factor got 6th rank with feasibility index 60 per cent. The last rank was for physical compatibility with least feasibility Index of 66 per cent. Overall feasibility index of Ergo stool for pruning and harvesting was 82.8%, indicating that it was acceptable to the workers.

Plate 4.3: Ergo stool for pruning and harvesting

Table 4.47: Feasibility index of Ergo stool for pruning and harvesting

(n=10)

Feasibility	Total	Feasibility index	Rank
Profitability	40	80	IV
Physical compatible	33	66	VII
Cultural compatible	38	76	V
Simplicity	50	100	I
Triability	40	80	IV
Grip fatigue	46	92	II
Physical stress factor	46	92	II
Work output factor	41	82	III
Tool factor	30	60	VI
Acceptability	50	100	I
	414	**(82.8%)**	

❑❑❑

Chapter 5

Discussion

This chapter presents the discussion regarding the findings of the study conducted under the three phases as follows:

Phase I : Work profile and working conditions of workers in grapes cultivation.

Phase II : Occupational risks of workers in grapes cultivation.

Phase III : Development an Ergo-solution for highly risk prone activity in grapes cultivation.

Phase I

5.1 Work profile of the workers

This section describes the remarkable results of the selected workers which revealed their personal and family variables. The background information of the selected population resources gives general information about a variety of topics. These are often considered to be general references, sources, meaning that they provide basic facts and knowledge that can be used as a foundation for one's research. A little time spent in background information save a tremendous amount of time, when searching in database and more subjects resources.

Socio-economic status is an individual's or group's position within a hierarchy social structure. Socio-economic status depends on a combination of variable, including occupation and place of residence. Sociologists often use socio economic status as a mean of predicting behaviour.

Results showed that majority of the workers (59.3%) in pooled sample belonged to old age and similar trend was also observed in three district i.e majority 50%, 75.0% and 66.6% respectively from Hisar, Sirsa, Fatehabad were in old aged. Cent per cent of the workers were male. Majority of the workers 44.4% obtained education up to primary school 38.8% upto senior secondary, which was the good indication for grapes farming. The grapes farming required educated man-power because a different type of pesticides and time bound farm work was required, this can be possible through educated worker. A thumbing majority of the workers 90.6% in total sample had temporary pattern of employment. It means that workers needs job/ work, so they were working in grapes farm on temporary basis. Majority of the workers were doing agricultural activities, so their source of income was from agricultural activities i.e. agriculture was basic source of financial support to the family. Two-third of the workers were having joint family with 4-5 members in the family. A huge majority of the workers 84.3% were married. Regarding annual income of owner from grape orchard, half of pooled sample was getting high income from grape orchards. It was found that majority of the owners were satisfied with income from orchards. Cent per cent of the workers belonged to general category. Regarding land holding most of the workers (62.5%) in pooled sample had less than 1 acre of land and 96.8% had buffalo. It indicates that basic priority of workers was agricultural and grape farming was done as side business. Cent per cent of the workers were doing grape farming from more than 9 years. It indicates that workers were well-experienced in their respective grapes cultivation activities. All workers in pooled sample had irrigated type of land and 81.2% in total sample had taken the land on lease for grape orchard. It indicates that maximum of the workers had less than 1 acre of own land holding that's they were working on lease type of land.

Years of grape farming : Overall data as well as district wise data indicate that cent per cent of the respondents were doing grape farming from more than 9 years. According to FAO reported that almost two-third (65.6%) of the respondents entered into grape farming during the period 1985-99, which corresponds to experience between 10-25 years in production of grapes. Respondents with experience of less than 5 years in grape production represented only around 12.5% of the entire sample (Shikamany, 2003).

Working hours in grape orchards

Data revealed that cent per cent of the workers were taking rest in between the work, and the length of work period/day was 4-8 hrs. A huge majority of the workers (83.3%) were taking resting period of 60-120 min and only 16.6% were taking rest period of more than 120 min. Mines *et al.* (2001) reported that individuals working full day under stressful conditions were more prone to accidents and injuries, thus occurred during work lead to other serious problem like crushing from farm equipment, acciendental slicing with hands labours tools and falling from ladders. Further, findings were in line with Mclean *et al.* (2001) who reported that person can actually work more productivity (and last much longer) with breaks between the works, and further showed that frequent breaks of 30 seconds to 10 minutes were beneficial. The benefits include increased performance and reduced fatigue to the eyes, lower back, neck and wrist, especially when breaks were taken at 20 minutes intervals rather than at 40 minutes interval. Similarly, Pheasante (2010) reported that people need to get up and walk around as much as possible to reduce health hazards as well as poorly designed workstation were increasing the length needed for rest breaks. Moore *et al.* (1991) showed that cumulated duration of exposure over the entire day was high, i.e. approximately 8 to 10 h per day over a 4-month period. Trimming the vines was performed after October and takes a team of 4-6 employees working 3 months, 8 hours a day.

Rest period are important to increase the performance and reduce the fatigue of workers, Hence, in grapes cultivation the workers were taking rest.

Mode of payment : Grapes workers were getting wages on weekly and contractual basis, there mode of fixing of wage was time of work and quantum of work. These finding were in line with Guarv *et al.* (2011) who reported that 44% of the respondent were getting wages of Rs. 100 per day, 31% respondents were getting in between Rs. 100 to 110 per day and other getting Rs.120 per day or more. Employees were working on grape farm on seasonal basis and not regular. Due to seasonal variation, the wage rates were continuously changing. Owners were giving salary as per minimum wages act, but along with direct wages, the grapes farm owner was providing transport facility to pick up workers from resident to the farm and drop back at evening, farm owners were also providing refreshment to the employees. In a grape harvesting season, the farm owner suppose to give two kg grapes per employee as a token of gift, the farm owner some time give Rs.120 wage to skilled and experienced women employees.

Level of tiredness and pain faced by grapes orchards workers

Cent per cent of the respondents in pooled sample and also in all district felt tiredness while doing activity. Further about one-third of the workers (31.1%) were feeling tired after working 3-4 hrs/day. In pooled sample, 43.7% each felt body pain / discomfort sometimes and always. Half of the respondents felt pain while doing work and other half felt after doing work. To overcome the pain and discomfort, about half of workers were taking rest. Risk factors of WMSDs include workplace activities, such as heavy load lifting, repetitive tasks, and awkward working postures. Demographic characteristics and psychosocial factors were also known to be important predictive variables (Haynes and Williams, 2008). William *et al.* (2007) have carried out a systematic review of psychometric evaluation of health related work outcomes measures for musculo-skeletal disorder. The purpose of the therapeutic interventions provided or prescribed by the physical therapy team were to either immediately impact (decrease) the worker's musculo-skeletal syndrome, and/or provide tools to address musculoskeletal dysfunction, and/or make a referral to another health care provider. Providing treatment to the workers at the vineyard clinics is a different model of care than the delivery of physical therapy in the traditional clinical setting. Education was also a key component to the intervention (Brummitt *et al.*, 2011).

Levels of problem faced by grape orchard worker in different district

Data represent the levels of problem in different grape orchards, Major problems which were physical problem, workplace problem, and problem due to high temperature, problem related to work, weather / climate, pesticides, economic problems, postural problem and injuries / insects bites. Physical problem include backache, asthma skin problems, pain in legs and pain in hands. Weighted mean score of postural problem of workers in three district (Hisar, Sirsa, Fatehabad) was 2.67 and rank I. Similarly, Villarejo *et al.* (2000) reported that heavy lifting, awkward body posturing, twisting and repetitive tasks of agricultural work lend themselves to the development of musculo-skeletal injuries that can pose acute problems and long-term disabilities among farm workers. Poorly designed tools, lack of training, and long work hours were the major factor contributing backaches and also leading to other chronic problems. Back and neck pain were the most common types of chronic pain workers experienced. Over 40 per cent of these workers left or changed jobs because of the pain they experienced (Estill and Tanaka 1998). Injuries/cuts, included cut, fracture, wound fall, insects /bites and scored overall score of 2.63

and rank II. These finding were in line with finding that injuries are almost three-times as likely to occur, when farm hires work long hours (more than 60 hrs/week) or on farms with large acreage (greater than 30 acres under tillage (Kidd *et al.* 1996). Workplace problems includes lack of cleanliness, lack of safety measures of workers, which got weighted mean score of 2.37 with rank III, however, Gaurav *et al.* (2011) also observed that 67% respondents were not aware about cleanness and hygiene. Regarding physical problems, which included backache due to prolong working, asthama, skin problems, pain in legs sneezing, breathlessness, eye irritation were observed with overall score (2.19) and rank IV. Ayub and Mittal (1989) found that repetitive movement (10-12/minutes), high pinch force (in tying the thread to wire and plant), frequent trunk flexion and extension motion, cutting finger (due to thread) and high eye strain were major problems in workers involved in tying activity.

The problems related to work include hard physical work, exertion due to over work, rude behaviour of owner, rude behaviour of co-worker and abusive language. The problems due to work scored overall weighted mean score of 2.13 and rank VI. Problems due to high temperature secured Overall weighted mean score of 2.12 with rank VII. Another problem was economics problems, which included delay in payments wages and low wages. The rank VIII was achieved by economic problems with overall score of 2.08. Bineford (2004) reported that some worksites, farm workers were not only putting in long shifts, six to seven days a week, but also working at an accelerated pace. Some respondents perceived that unreasonable productivity targets, piece rate wage systems, and pressure from employers and supervisors intensified the production process to an extent that was increasing their risk of workplace injury. Age safe (1992) data suggest that 48.9 per cent of all reported injuries in horticultural specialties (including nursery) were sprains and strains. Overexertion as a cause of injuries for this area was cited for 30.2 per cent of the reported injuries.

Area of grape orchards

Results represent the area of different grapes orchards. The maximum area of grape orchard was in Hisar district i.e. 45,760 metre square followed by Sirsa 44,000 sq mt. and Fatehabad 39,600 sq mt. and least areas was also in Hisar district 36,300 sqmt. According to FAO reported that around two-fifth (40.7%) of the respondents operated an area ranging from 1 to 2 hectares for grape farming.

Respondents managing an area of 2 to 5 hectares under grapes comprised slightly less than a one-third (31.2%) of the sample. The representation of growers with a hectare or lower area under grapes cultivation was close to a one-sixth (15.6%) of the entire sample (Shikhamany, 2003).

Varieties of grapes cultivated

Results represent that mainly three varieties of grapes were cultivated in the grape orchard. In pooled sample, majority of the respondents (66.6%) were cultivating perlette variety and 16.6%, each were cultivating Anab-e-shabi and Thomson seedless. In Hisar and Fatehabad, perlette variety had been cultivated in all the orchards, Whereas in Sirsa Anab-e-shabi and the other 50% were cultivated Thomson variety cultivated by 50% each. According to NHB Horticultural data base reported that in India, main grape varieties and their proportion was Anab-e shabi (15%), perlette variety (5%), Thomson seedless (55%). These varieties growing under north-India conditions. On the other side, cent per cent of the respondents were using telephone system for training and greatest advantage with this system was the protection of bunches from direct sunlight. Raising method was used for cultivation of grapes, because grape can be easily propagated by hardwood stem cutting prepared from mature, one year old cane removed from pruning wood (Shikhamany, 2003).

Training profile of the owners in orchards : Cent per cent of the owner had acquired training for grapes cultivation. Mode of training of cent per cent was lecture and one-third had attended workshop for grapes cultivation. In Hisar and Fatehabad, all the grape orchard owner had attended the lecture and 50% had attended workshop. The employees of grape orchards received training regularly. The training sessions include trimming technique, application of phytosanitary products and manual handling of loads.

Posture used and risk prone areas of body in activities performed by the workers in grapes cultivation

Land preparation activity was done by 26 workers, during the month of December through use of *spade*. The bending posture was adopted, by workers and risk prone areas of body were lower back, neck, mid back. Pruning was done by 30 workers during the month of January through use of scateur. James *et al.* (2010) reported that worker stands facing the vine , reaches in with the dominant hand gripping shears, and cuts to be pruned. Highly repetitive and forceful hand closures to make cuts (36-46) per minute and impact imparted to hand as the handles of the pruners close at the end of each cut, pressure on the dominant hand from pruners handles and non-dominant hand from branches being gripped and cut. During pruning hand devices, such as knives, scissors and saws were used. Self propelled machines and high-tech pruning methods were only suitable for new wine grape plantation.

Irrigation activity was done by 6 workers during the month of February to April through tubewells. Vineyards border irrigation needs much more water, manpower, time and as a result higher cost. Although higher capital demand at the beginning, the water and energy costs of drip irrigation were approximately 20-40% lower. Manuring activity was done by 8 workers during the month of February to April through tractor and manual spray pumps. Surface spreader machine was much more efficient and saves about 30% fertilizer, although the time requirement was little higher. Harvesting activity was done by 29 workers during the month of May/June either manually or use of shaving blade. All vineyards in the region were harvested manually.

Environmental conditions of grapes orchards

A comfortable work environment is very important to perform the task easily. The health and safety of grape orchards workers. A compatible work environment is very important to perform the task easily and to increase the productivity with reduced numbers of errors. Hence, the grape orchard environment was assessed with environmental parameters which included temperature, relative humidity, light, CO_2 and SPM (suspended particules matters).

Temperature : The recommended value of the temperature was 20- 24^{0}C. In winter season, mean temperature in the grape orchards was less than recommended value. During summer season, mean temperature in the grapes orchards was more than recommended value in three district, which was resulting in discomfort.

Relative humidity (%) : Recommended value of the relative humidity is (40-60%). In both seasons, winter and summer level of relative humidity was more than recommended value. Bakshi (1997) reported that the comfortable limit of humidity is 40-50% for winter and 40-60% for summer, whereas relative less than 30% is undesirable.

Light (Lux) : Recommended value of the illumination level is 500-1000 lux. In winter season, level of light in the grape orchards was less than recommended value. In summer season, level of light in the grape orchards was more than recommended value. In summer, the light level of convenient as compared to winters.

Carbon dioxide (CO_2 ppm) : Recommended value of CO_2 is 350-450 ppm. In winter and summer, both seasons in the grape orchards CO_2 was less than recommended value. Hence, it was safe to work.

SPM (Suspended particules matters ppm) : Recommended value of SPM is 140 ug/m^3. In winter season, the value of SPM was less than recommended value in the grape orchards. In summer seasons, SPM was higher in the grape orchards than recommended value. So, it was too safe to work in winter season but not safe in summer season. The environmental conditions for workers in grapes cultivation were comfortable in winter season as compared to summer season.

Effect of Heat stress in grape orchards

Results regarding heat stress in grape orchard depicts that during winter season in Hisar, Sirsa and Fatehabad environment was comfortable with heat stress score 79, 71 and 71, respectively, which indicates that there was no problem of heat stress in the winter season. In summer season, heat stress in grape orchard of Hisar was 113, and that of Sirsa and Fatehabad was 92 & 94, respectively, this depicted that extreme caution was required by the workers, as heat stress could result in sun stroke, heat cramps and heat exhaustion possible with prolonged physical activity. Portier *et al.* (2012) reported that in the summer time, when air temperatures are already high, elevated humidity levels can increase people's risk for heat-related illnesses. Symptoms of overheating include heat cramps, heat fainting, heat exhaustion and heat stroke. During hot summer months, the combination of hot air temperatures and high humidity levels put farm workers in danger of heat-related illnesses, such as heat stroke and heat exhaustion. Further as per report of U.S., the annualized rate of reported heat-related deaths among crop workers was 2.36 deaths per 100,000 workers (Luginbuhl *et al.,* 2013).

Phase II

5.2 Experiment work

Health status of the grape orchards workers selected for the experimental work

For experiment work, those workers were selected who had high average to good health status. The mean body temperature, blood pressure, pulse rate of the grape orchards workers were within the normal range. Aerobic capacity (Vo_2max) of the grape orchards workers was average.

Physical characteristics and health status of grape orchards workers

The experiment was conducted on 15 healthy grape orchard workers. The selected workers were having an average height of 159.9±8.8 cm with weight 64.2±4.7kg. Body temperature, blood pressure (mm/hg), pulse pressure (mm/hg), pulse rate (bpm), body mass index, body composition of all the workers were within normal range. The workers had mesomorph body type indicating a good body built.

Occupational risks of workers in grapes cultivation

Occupational risks of workers in grapes cultivation was assessed through following parameters:

i) Physical load
ii) Biomechanical stress
iii) Physiological load
iv) Time load

I Physical load : Physical load was assessed through:

1. Load rating
2. Workload assessment through Key Indicator Method (KIM)
3. Physical load

Occupational risks of grape orchard workers as per load rating

Occupational risk in various activities of grapes cultivation as per load rating revealed that land preparation was considered very heavy activity. Pruning and harvesting were heavy activity. Whereas, handling and transportation were considered as moderately heavy, manuring, irrigation, plant protection were considered as low activity. Conclusively land preparation was very heavy activity as per load rating.

Assessment of physical workload during manual handling operation in grapes cultivation through Key Indicator Method (KIM)

Key indicator method or assessing physical workload during manual handling operation in grapes cultivation reflected that in land preparation activity. The risk score for both hands depicts the increased load situation, physical overload, hence re-design of workplace should be done. In Pruning, KIM score for left hand indicate increased load situation, physical overload may possible, hence re-designing of workplace should be reviewed. KIM score for right hand highlighted that workplace re-design is necessary, as there was high load situation; physical overload is likely to appear. A handling and transportation and land preparation activity both hand depicts the increased load situation, physical overload, so re-designing of workplace should be reviewed. In manuring, irrigation and plant protection KIM Score for left hand indicate depicts moderate load situation, physical overload also possible for less resilient persons, so re-design of workplace should be reviewed. KIM score for right hand indicate increased load situation, physical overload also possible for normally resilient persons. Re-design of workplace should be reviewed.

On the basis of KIM score, right hand in pruning activity and both the hands in harvesting and handling and transportation were highly risk prone as there was high load situation, physical overload is likely to appear and hence, workplace re-design is necessary.

However, Wakula *et al.* (2000) reported that pruning grapevines and wine harvesting involves a combination of dynamic and sensorimotric work, and also a high incidence of ergonomically undesirable posture of the trunk, the upper limbs and the head irrespective of cutting tools used.

Occupational risks of grapes workers as per physical load

Occupational risk in various activities of grapes workers in terms of physical load was to assessed in terms of weight of the load, distance carried, height lifted and physical load score. Physical load was the highest in manuring followed by irrigation and pruning. Physical load factor score of other activities in decreasing order was plant protection, harvesting, handling and transportation. Land preparation got least physical load factor score.

Conclusively in terms of physical load manuring, irrigation and pruning were most risky activity and least risky activity was land preparation.

Total Occupational risks of grape orchards workers in terms of physical load

The total occupational risk of workers in terms of physical load was assessed on the basis of load rating, workload assessment through key indicator method and physical load rating. The physical load was maximum for pruning followed by harvesting. The total physical load of other activities in descending order was handling and transportation, land preparation, manuring and irrigation. The least total physical load was for plant protection.

On the basis of total physical load in the grapes cultivation pruning activity had highest occupational risk.

Biomechanical risks of the workers while working in grapes cultivation

Biomechanical stress was assessed through:

- Grip fatigue
- Angle of deviation
- RULA score (Mc Atamney, Corlett. 1993)
- OWAS (Karhu *et al.* 1997)
- Posture load score
- Repetitive strain exertion
- Musculo-skeletal discomfort

Grip fatigue of hands before and after performing grapes cultivation activities

Mainly daily functions and sporting events require high levels of the flexor musculature of the forearms and hands. During gripping activities the muscles of the flexor mechanism in the hand and forearm create grip strength while the extensors of the forearm stabilize the wrist. Handheld grip strength dynamometry was used to measure the muscular force generated by flexor mechanism of the hand and forearm. Grip strength was moderately correlated with overall body strength. Grip strength gives the measures of fatigue. As the strength decreases, the fatigue increases Smith *et al.* (2006). The grip fatigue in pruning activity was the highest. Grip fatigue for right hand was 30.9% and grip fatigue of the left hand was 31.0%. In land preparation was 30% and 29.8%, respectively for right hand and left hand respectively. In manuring activity, the grip fatigue of right hand was 20% and left hand was 20.3% followed by irrigation activity, with grip fatigue of right hand 20% and left hand was 19.7%. However, in harvesting activity, the grip fatigue of right hand was 14.9% and for left hand, it was 15.1% and in handling and transportation, the grip fatigue of right hand was 13.7% and left hand was 14.9%. The least grip fatigue for right hand was 15.7% and for left hand was 6.7.

The grip fatigue was maximum in pruning activity for both the hands. Hence, depicting that in pruning activity, there was fatigue in the hands of the workers. Fry *et al.* (2006) also found correlation between grip strength and performance. Performance of highly repetitive task was associated with force dependent decline in grip strength and increased fiberogenic related problem, this change was attenuated least short term, by anti-inflammatory treatment. Data reflect that pruning was an activity of cutting extra leaves and weeds to make the space between plants. It was highly prolonged activity in stooping posture and frequent lifting and hand intensive task.

Angle of deviation of cervical and lumber region before and after performing grapes cultivation activities

Excessive spinal deviation is considered very harmful for the workers, as it exposes them to risk of development of low back pain. The normal spinal angle of worker in grapes cultivation was 181.9^0.

The maximum deviation in the spinal angle was in land preparation, followed by manuring, irrigation, plant protection and pruning and least was for harvesting, handling and transportation. More the deviation in spinal angle, higher are the chances of musculo-skeletal discomfort in neck, shoulder and back. Hence, in the land preparation activity, MSD will be more. Similarly, Mayer (2010) reported that repetitive washing tasks with lifting and carrying of beds and moving materials in awkward posture were found to be big problems. The spinal deviation in bed washing was highest and was at 21^0 of flexion, which were causing the back injuries with prolonged working. Problem in neck, shoulder and low back were found higher because workers were doing full day job in stooping posture with forward bending of neck, shoulder and back. Similarly, Grandejean (1998) reported that constrained body posture and deviation from neutral posture were the most frequent of static muscular effort. Borah (2009) conducted a study on women labour engaged in cashewnut industry. The angle of deviation of spinal curve was above the normal range while performing the activity, which was the high risk factor for musculo-skeletal disorders. Similarly, Meyers *et al.* (2000) reported that pruning, weeding and labelling were particularly stressful due to prolonged extreme flexion of the trunk ($>60^0$).

Rapid upper limb assessment (RULA) score of workers during the activities of grapes cultivation

The rapid upper limb assessment (RULA) a posture targeting method was used for estimating the risk of the work related upper limb disorder. A RULA assessment gives a quick and systematic assessment of postural risk at a work (Metamney and Nigel, 1993). RULA score was used to assess the discomfort in upper limb during the various activities in grapes cultivation. Rapid upper limb assessment of workers in the various activities of grapes cultivation emphasized that, investigate and change immediately was required in pruning, harvesting and land preparation whereas investigate further and change soon was required in manuring, irrigation, plant protection and handling and transportation.

Louhevaara and Suurnakki (1992) also reported that working postures cause static load on the musculo-skeletal system. Static work results in low efficiency of work. The reduction of the static load caused by poor work posture was one of the main measures for correcting the situation, where there was constant increased prevalence of back disorder and musculo-skeletal complaints. Similarly, Mayer (2005) reported that high risk factors in pruning in vineyard were prolonged trunk flexion (up to 90 degree of trunk flexion, 50 degree flexion of lumber), awkward posture of both wrist, primarily flexion and ulnar deviation and continuous stress by handling/gripping sheers. Similarly Chapman *et al.* (2004) also reported that fresh market vegetable farmer produce vegetables that are sold at markets and roadside stands are exposed to high stress on the low back during lifting of containers filled with vegetables.

Ovako working posture assessments (OWAS)

The workers of grapes cultivation activities work in repetitive movements and can cause musculo-skeletal disorders. Work involving repetitive movement, body stretches more bending while working can result in health problems due to overuse of certain muscles. The OWAS score of the grape workers was assessed to know the posture while working. Ovako working posture assessments (OWAS) of majority of workers in the various activities of grapes cultivation heighlighted that corrective measures immediately was required in pruning and harvesting activity, whereas corrective measures as soon as possible was required in land preparation, manuring, irrigation and handling and transportation, However, in plant protection, activity required no corrective measures, Kumar (2000) reported that any force exertion, repetition of activities or assuming one particular posture for prolonged period imposes stress on human system and should be changed. Similarly, Dalal *et al.* (2010) found that the respondents reported having pain experiences in lower back (54%), neck (49%) and upper back (48%). There was a decrease in OWAS score to one number after minor engineering changes introduce. In same results were found that major risk factor in nursery was awkward posture. A worker experiences upto 60 degrees of trunk flexion, when placing cans on the trailer (on an average of 40 degrees of peak lumbar flexion), where upto 90 degrees of shoulder flexion, when placing cans on the trailer. This represents high biomechanical stress on the shoulder. Similarly, Weeding in plant nurseries produces severe stooped postures for extended times that may lead to LBP (low back pain) (NIOSH, 2001).

Hence, on the basis of OWAS as well as RULA, it can be concluded that posture used in pruning and harvesting is not suitable and it needs to be changed instantly.

Occupational risks in various activities of the grapes orchards workers in terms of the posture load

Occupational risks in various activities of the grapes cultivation in terms of the posture load means echlon of risk in the use of posture and the body parts involved. The occupational risk in the activities of grapes cultivation in terms of posture load factor was highest in pruning activity followed by land preparation, manuring, harvesting and plant protection. The least posture load score was for irrigation and handling and transportation. Higher the posture load, more the risk involved in the activity. So, pruning activity was the risky, as it secured highest score in posture load pruning, land preparation, manuring, harvesting and plant protection. Janowitz *et al.* (2000) reported that sustained awkward postures occur, when harvesting grapes with trunk flexion angles between 20^0 and 45^0 being held repeatedly for upto 25 seconds at a time.

This indicates that posture in pruning was most risky i.e. it was hard to maintain the posture during the pruning activity. According to posture load, handling and transportation was easy to maintain. Meyers *et al.* (1999) reported that that awkward postures in the harvest task include moderate shoulder flexion and mild shoulder abduction; frequently severe forearm pronation and occasionally severe forearm supination , moderate-to-severe ulnar deviation and virtually no radial deviation of the wrist. Trunk flexion ranges from mild during cutting to severe during stooping to gather fallen grapes. Neck flexion is occasionally severe, but often the neck is in extension of more than 5 degrees. Workers were often observed in a forward head position of trunk flexion with simultaneous neck extension. Additional risk factors include: constant ulnar deviation of dominant hand, very high energy cost, pressure on hand from poorly-designed knife handle cuts on non-dominant hand from knife and long carries with bins held overhead for some work methods. Similarly, finding reported that Hokm Abadi (2012) showed that about 56% of the postures related to the right side of the body were classified as medium level and about 30% as high level. But about 76% of the postures related to the left side of the body were classified as medium level and about 16% as high level using REBA. Furthermore, the results of the study by Miri (2008) showed that about 42% of the postures related to the right of the body were classified as medium level and about 38% as high level, but about 46% of the postures related to the left side of the body were classified as medium level and about 40% as high level using REBA. Nevertheless, in the research by Frvarsh, about 27.6% of the postures were classified as action level 2-3 (medium risk high risk) using REBA.

Repetitive Strain (exertion) - The repetitive strain in grape cultivation was assessed through:

- ART tool (assessment of repetitive task)Health and safety executive (2010)

- Strain Index (7- Point continuum) (Moore and Garg, 1995)
- Repetitive Strain load score

Assessment of repetitive tasks of the upper limbs in grapes cultivation activities, showed that in pruning, the exposure score for both hands indicates that further investigation was required urgently in the pruning task in harvesting and land preparation. The exposure score for both hands generates that further investigation was required in the task. In handling and transportation activity, further investigation was required urgently. In manuring, irrigation and plant protection depicts that consider individual circumstance (Both hands).

As per the assessment of repetitive tasks (ART), in grape cultivation pruning was risky activity and further investigation was required immediately.

Grapes orchards workers were exposed to ergonomics problems in their routine works. The ergonomics problems were unavoidable, such as awkward posture, repetitive movements. Weight handling, force in pulling and pushing activities. Montomoli *et al.* (2010) reported that the ergonomic analysis using OCRA check-lists and NIOSH method to study the spine and upper limb work-related musculo-skeletal disorders in grape orchards. Each tasks analyzed showed a high risk of biomechanical overload of the upper extremities. Regarding the manual material handling in the grape-harvest and of risk changed from yellow to green.

Strain index in various activities of grapes cultivation

The strain index in the activities of grapes cultivation was used to assess hazardous activity, the activities except manuring and irrigation were hazardous. The strain index of various activities was pruning (175.5) harvesting (54.0) handling and transportation (54.0), land preparation (6.8), whereas, manuring, and irrigation got least strain index (3.4) and its level of hazard was uncertain.

As per strain index of the activities of grapes cultivation, pruning harvesting and handling transportation were hazardous activities, but among these pruning score the highest score. The finding were in line with Janowitz *et al.* (2000) who reported that vineyard workers face high stress on the hands during pruning of the grapevines under highly repetitive conditions (8 to 10 week period) of intense and fast-paced work.

Repetitive strain load score of workers in grapes cultivation

The activities of grapes cultivation were repetitive in nature hence resulted in occupational risk in terms of repetitive strain. The highest repetitive strain load factor was in pruning and handling and transportation followed by irrigation, harvesting, land preparation and manuring. The lowest repetitive strain load was for plant protection. Kumar (2000) reported that any force excretions, repetition of activities are assuming one particular posture for prolonged period imposes stress on human system and should be changed. Work involving repetitive movements, body stretches more bending while working can result in health problems due to overuse of certain muscles. Similarly, Fulmer *et al.* (2002) reported that cutting grapes from the vine during harvesting also creates high stress on the hands and wrist through high repetition (25 to 50 cuts per minute), which includes contact stress between the knife and hand. Fruit harvesters, such as apples, pears, oranges and other tree-born fruit are exposed to different high shoulder stresses during three tasks, reaching upto pick fruit above shoulder height, carrying bags placing loads on shoulder through strap use, and carrying the ladder from tree to tree. Similarly, Nursery farmers face several hand-wrist risk factors, when performing tasks, such as pruning, plant propagation, and weeding. These risk factors include poor wrist postures, repetitive gripping, high-hand forces specifically pinch forces, and contact stresses. Crop harvesting has several risk factors for hand and wrist disorders high hand forces, when pulling vegetables or fruit off plants or using cutting tools to remove crops, high hand repetitions greater than 15 times per minute, static loads when cutting off crops, and awkward wrist postures (Meyers *et al.*, 2000).

Overall Repetitive strains (exertions) in grapes cultivation

Overall Repetitive strain score was assessed in terms of ART tool, strain index, repetitive strain index score. Overall maximum repetitive strain (exertion) was the highest in pruning 239.5, followed by harvesting, handling and transportation, land preparation, and manuring, irrigation plant protection were least hazardous activity. Hence, pruning was risky activity in terms of repetitive strain.

Musculoskeletal Disorders /discomforts of grape workers

Musculo-skeletal disorders of grapes workers were assessed as per:

- Human body map
- VAD scale
- Nordic scale

Intensity of musculo-skeletal pain in workers of grapes orchards as per human body map

Intensity of discomfort of grape workers was recorded after performing the grapes cultivation activities on a five-point scale as per human body map as indicating very severe pain, severe pain, moderate pain, mild pain and very mild pain. The intensity of musculo-skeletal discomfort was highest in buttocks, wrist/hand and neck followed by ankle/feet, shoulder, elbow upper, arm, and muscles lower arm and eye. Least discomfort was in ankle/feet, calf muscle. Risk factors for upper–extremity musculo-skeletal disorder include biomechanical factors (force, repetition, posture and psycho-social factor (job stress) as stated by Morse *et al.* (2007). Aweto (2015) reported that the low back pain was the most common area of discomfort, followed by the shoulder and then the neck. Similarly, Kittusamy and Ambrose (2004) reported that 70% of operators of farm equipment suffer in one or more body regions-neck, upper back, lower back, shoulders, forearm/elbow, wrist and hand, hip, knee, and ankle/foot. Menzal (2007) also has brought out the role of psycho-social factor (job strain, social support at work, and job dissatisfaction) in musculo-skeletal disorder. He suggested measures for reducing the incidence, of musculo-skeletal disorder and addressing psycho-social risk factor. Similarly, Youakim (2006) reported that tasks, such as pruning and harvesting cause repeated stressing of hands and wrists, therefore MSDs were very common during the pruning and harvesting seasons. Similarly, Davis and Kotowki (2007) reported that during harvest, lower back, hands, and shoulders are stressed multiple times, when trays or buckets filled with grapes are lifted from the ground and carried to the trailer. Low frequency vibrations from machines also cause lower back problems.

Visual analogous Discomfort (VAD) in grapes cultivation

The visual analogue discomfort rating scale was used to flaunt the amount of discomfort in different activities performed by grape workers. On the basis of observations land preparation, pruning and harvesting activity were found to be very vigorous activity. However, handling and transportation were considered as moderate activity. Further manuring, irrigation and plant protection were treated as mild activities.

Land preparation, pruning and harvesting activities were considered as the vigorous activities by the workers in grapes cultivation, hence were risky as per VAD score. Work safe Bc (2009) reported that half of the claims of injuries in the agriculture industry were related to ornamental nurseries, greenhouses, and farm labour supply or services.

Nordic musculo-skeletal questionnaire assessment for grapes cultivation

The musculo-skeletal problems and the pain perceived during MMH (Manual material handling) activity with different methods were determined by administering standardized Nordic Questionnaire. All the selected workers had given their response, which were analyzed. In the study an appreciable number also reported musculo-skeletal problems, such as backache and pain in upper limbs. Nordic musculo-skeletal questionnaire assessment results revealed that maximum in handling and transportation followed by pruning, harvesting and plant protection and land preparation maximum prevalence of musculo-skeletal problems higher in elbows and upper back, one or both ankle in neck, wrist, and lower back.

Meyers *et al.* (2000) reported that main risk factor for upper back and neck MSDs in farming was severe neck flexion or extension. Many farming tasks produce severe neck flexion that was of ten-times held for long durations, such as: harvesting of plants on the ground, weeding fields, and pruning plants. Neck extension occurs the farm workers do tasks above their heads in tasks, such as fruit harvesting and wine grape pruning. Long durations of the neck being twisted during driving of tractors in the field. Duraj *et al.* (2000) reported that high stress on the shoulders, when grape harvesters carried full bins of grapes above their head the fields and the truck or emptying bins into large collector (bulk) bins. Similarly, Larson and Hannihen (1995) unfolded that excessive musculo-skeletal stress at work, specially with static load, as it plays a major role in low back pain, neck and shoulder disorders. Electromyography recording during working conditions has been used to quantify muscular stresses, allowing better designing of work environment to reduce low back pain and neck shoulder tensions.

Wrists and neck, shoulder, lower arms and upper back was the frequently used body part in grapes cultivation activities. Pain felt in the other body parts were 'buttocks due to adoption of poor posture for prolonged period. YU and Wong (1996) found in study that frequent users of VDU (Visual display units) had significantly more musculo-skeletal problems in the neck and shoulder regions than infrequent users. Individual musculo-skeletal complaints were associated with various risks factors including personal attributes, working postures, repetitive movements and work station design. Back, neck and shoulder problems were more related to unfavourable working postures, arm, hand and wrist problems were more affected by repetitive movements. Similarly, Hildebrandt *et al.* (1995) reported that 75% of farm workers reported experiencing musculo-skeletal symptoms during the previous 12 months.

Musculo-skeletal discomforts in grapes cultivation

The musculo-skeletal discomfort was assessed on the basis of Human body map, VAD scale, and Nordic scale. The musculo-skeletal discomfort was maximum for harvesting total score followed by land preparation. The MSDs in other activities in descending order was pruning, handling transportation, and plant protection. The least of total musculo-skeletal discomfort was for manuring and irrigation.

On the basis of musculo-skeletal discomfort in the grapes cultivation harvesting activity had highest MSD. Musculo-skeletal disorders of the hips, knees, legs, ankles, and feet occurred when workers stand and walk on uneven or muddy ground, carrying heavy loads, squatting for long periods or standing on hard ground. Osteoarthritis is also a common hip problem among vineyard and farm workers. Takula (2008) reported that most musculo-skeletal disorders were cumulative disorders resulting from repeated exposures high or low intensity loads over a long period of time.

Overall Occupational risks of workers in grapes cultivation in terms of Biomechanical stress

Overall occupational risks in grapes cultivation in terms of biomechanical stress was assessed on the basis of grip fatigue, RULA, OWAS, posture load score, repetitive strain exertion and musculo-skeletal discomfort.

The occupational risks in terms of biomechanical stress was highest in pruning followed by harvesting. The other activities in descending order were handling and transportation, land preparation, manuring, and irrigation. The plant protection activity was least risky in terms of biomechanical stress.

Physiological Load

The physiological load in grapes cultivation was assessed through:

- Heart rate (bpm)
- Oxygen consumption (l/min)
- Energy expenditure(KJ/min)
- Perceived Exertion
- Physiological load factor

According to Kroemer *et al.* (1997) the heart rate and oxygen consumption have a linear relationship. They found that the relationship may change within one person with training, and it differs from individual to another. They inferred that heart rate measurement could be substituted for measurement of metabolic processes, particularly for oxygen consumption, since it could be performed easily. Heart rate, energy expenditure and oxygen consumption rate of worker in grapes cultivation. Regarding physiological stress in working conditions maximum heart rate (bpm) was in land preparation followed by pruning and harvesting. Energy expenditure was also higher in land preparation activity, followed by pruning and harvesting. Oxygen consumption rate was also higher in land preparation activity followed by pruning (0.62 $l.min^{-1}$) and followed by harvesting (0.51 $l.min^{-1}$) with the increase in heart rate and energy expenditure the risk factor also increase, so pruning was not risky activity. Similarly, Sanders and McCormick (1993) stated that the linear relationship between heart rate and oxygen consumption is different for different people. So, they suggested calibration of each person to determine the relationship between heart rate and oxygen consumption. They reported that heart rate is the best used as a predictor of oxygen consumption, when moderate to heavy work is performed. They also stated that heart rate continuously sampled over a work day or task, is useful as a general indicator of physiological stress without reference to oxygen consumption or energy expenditure.

Association of age, frequency of rest period and BMI between Physiological parameter

There was significant association of heart rate, oxygen consumption, and energy expenditure with age was found in land preparation, pruning, harvesting and non-significant association of heart rate, oxygen consumption, energy expenditure with age was found in manuring and irrigation activity. Similarly, there was significant association of heart rate, oxygen consumption, and energy expenditure with frequency of rest period in land preparation, pruning, harvesting and non-significant association of heart rate, oxygen consumption, energy expenditure with frequency of rest period was found in manuring and irrigation activity. On the other hand, a significant correlation existed between the heart rate, oxygen consumption, energy expenditure and BMI, in all the activities of land preparation viz, land preparation, pruning, manuring, irrigation and harvesting. Work related musculo-skeletal disorder did not affect ¾ of the participants' daily activities and 2/3 of the participants reported that symptoms had no effect on their job efficiency. The majority of the participants had a gradual onset of musculo-skeletal disorder while 7.9% reported a sudden onset.

Rating of perceived exertion (RPE) experienced by the grapes cultivation workers

Borg rating of perceived exertion (RPE) was used to measure the intensity of grapes cultivation activities. On the basis of responses regarding RPE obtained by the grapes cultivation activities, it was concluded that RPE for pruning activity was very heavy exertion followed by land preparation, harvesting which were considered as heavy exertion and irrigation and handling transportation were activity with moderate exertion followed by plant protection, manuring activity with very light exertion. On the basis of RPE pruning was perceived as very heavy exertion activity, hence it was most risk activity.

Occupational risks of grape workers in terms of physiological load

Occupational risks in various activities of grapes cultivation in terms of physiological load. The physiological load factor was highest in the pruning followed by irrigation. The physiological load factor of other activities in descending order was for handling and transportation, taken over by harvesting, land preparation activity. Manuring and plant protection were least risky. Activity with higher physiological load was risky with increase in physiological load risk factor also increase. Hence, in terms of physiological load factor, pruning and irrigation were most risky activities.

Overall occupational risks of workers in grapes cultivation in terms of physiological load

Overall occupational risks of workers was assessed on the basis heart rate, energy expenditure and RPE scale and physiological load. According to physiological load, pruning got Ist rank and harvesting secured IInd rank. Land preparation obtained IIIrd rank followed by irrigation, which got IVth rank. Handling and transportation got V rank followed by manuring. Last rank was secured by plant protection.

Occupational risks of workers in terms of time load factor

Occupational risks of workers in terms of time load indicates time taken to complete the activity. Occupational risks in various activities of grapes cultivation in terms of time load factor was highest in pruning followed by harvesting. Time load factor of other activities in descending order was handling and transportation followed by plant protection, land preparation, irrigation. Higher time load indicate more time consumption to complete the activity, more time consumption will cause fatigue and more MSDs. Conclusively pruning activity was most hazardous activity in terms of time load, and lowest hazardous activity in terms of time load was plant protection. Carruth (2002) reported that working on the farms and the size of the farm has found to influence injury rates. Stueland (1997) reported that injuries increased as hours worked per week increased, about 3% for every hour worked. Injuries are almost three-times as likely to occur, when farm hires work long hours (more than 60 hrs/week) or on farms with large acreage (greater than 30 acres under tillage. Similarly, Janowitz *et al.* (2000) reported that vineyard workers face high stress on the hands during pruning of the grapevines under highly repetitive conditions (8 to 10 week period of intense and fast-paced work.

Overall occupational risks of workers in grapes cultivation

Overall occupational risks of workers in grapes cultivation was assessed on the basis of physical load, biomechanical stress, physiological stress, time load factor.

Overall maximum occupational risks of the grapes workers were highest in pruning followed by harvesting, handling and transportation and land preparation. The medium risky activities were manuring and irrigation. Plant protection was least risky activity.

5.3 Phase III

Development of an ergo solution for highly risk prone activity in grapes cultivation

On the basis of results of Phase II pruning and harvesting of the grapes were identified as the highly risk prone activities. So, available technology used by them was modified and an ergo solution was developed for these two activities.

Available technology i.e. scateurs was modified by adding the materials on grip /handle, and an ergo-solution for pruning and harvesting activity was developed and was named as ergo stool for pruning and harvesting. The acceptability and feasibility of ergo stool for pruning and harvesting was assessed.

Acceptability of improved grape pruning secateurs by orchard workers

The Acceptability of improved grape pruning secateurs was obtained from 10 grape orchard workers. Ranks were computed on the basis of mean scores obtained in the following parameters:

"Liked the appearance", 'pruning can be done in proper way' and 'multipurpose in usage secured' 1st rank. 'Increased coverage area under harvesting and 'pruning got 2nd rank. 'Comfortable size and shape of the grip while using' secured 3rd rank followed by the 'good replacement of the existing tool and technique'. 'Able to maintained comfortable body posture in standing position while using modified scateurs,' 'minimum twisting of trunk while doing the activity'. 'However, restrained movements of hands' secured mean score of 2.2 and last rank was secured by 'coarse hard and unsuitable grip' Janowitz *et al.* (2000) reported that utilizing smaller and lighter weight bins during the harvesting of wine grapes. These intervention tubs reduced the weight by 12%, reduced the horizontal moment of arm by reducing the depth of the bins, included handles, and had smooth bottoms that reduced the sliding forces by 32%. Based on the NIOSH lifting guide, the lifting risk was reduced by 23% for the intervention tubs versus traditional tubs with virtually no difference in productivity. Similarly, Duraj *et al.* (2000) found that the impact of a tub cart for the transferring of tubs between fields and truck or collector bin. Workers had better postures i.e. back flexion, and knees only slightly bent.

Feasibility of improved grape pruning secateurs

Feasibility of improved grape pruning secateurs in terms of profitability, physical compatibility, cultural compatibility, simplicity, traibility, grip fatigue, physical stress factor, work output, tool factor and acceptability.

In profitability that (80%) of the respondents considered improved grape pruning secateurs as most feasible, 10 per cent each considered it as feasible and somewhat feasible, technology for pruning of grapes in terms of profitability. In terms of physical compatibility majority of the respondents (90%) considered improved grape pruning secateurs as most feasible and remaining 10 per cent considered it as feasible technology. Regarding cultural compatibility 40% of the respondents considered improved grape pruning secateurs as most feasible, 60 per cent respondents considered it as feasible technology. In simplicity most of the respondents (80%) considered the improved grape pruning secateurs as most simple and rest of 20 per cent respondents considered it as simple technology for grape pruning. Half of the respondents considered it as most feasible followed by 30 per cent who considered it as feasible and 20 per cent considered it as somewhat feasible technology for triability. Eighty per cent of the respondents considered grape pruning secateurs it as most feasible, 10 per cent each considered it as feasible and somewhat feasible, grape pruning technology in regards to grip fatigue. Physical stress factor 40% of the respondents considered grape pruning secateurs it as most feasible, half of the respondents considered it as feasible and 10 per cent respondents considered it somewhat feasible technology for grape pruning.

Majority of the respondents (80%) considered improved grape pruning secateurs as most feasible and remaining 20 per cent respondents considered it as feasible technology for work output. Most of the respondents, 70% considered grape pruning secateurs as most feasible, and rest of (30%) respondents considered it feasible technology for pruning of grapes in terms of tool factor. Cent per cent of respondents considered grape pruning secateurs as most feasible, technology for grape pruning.

Overall feasibility of improved grape pruning secateurs was maximum (71.0%) as most feasible followed by feasible (24%) & somewhat feasible (5%), as grape pruning technology. Newenhouse *et al.* (2005) reported that electric pruners were maximum potential to save labour about 20% and they cost approximately $1,500. Similarly, NIOSH (2001) found that simple ergonomics solutions for farm workers are extended lifting tool for picking up plants in containers that reduces severe forward flexion of the trunk and eliminates pinch grip of hand, weeding stand for nursery workers who weed potted plants that reduces prolonged stooping and neck flexion, standard picking bins with handles that reduces the weight lifted and provides better grasping surface, a lighter-weight berry harvesting rake that reduces prolonged stoop postures and reduces weight of the rake, a packing shed layout that adjusts the work area to the farm worker and increases productivity.

Development of ergo solution i.e. Ergo stool for pruning and harvesting Acceptability of Ergo stool for pruning and harvesting by grape orchard workers.

The Acceptability of Ergo stool for pruning and harvesting was obtained from 10 grape orchard workers. Ranks were computed on the basis of mean score. "liked the appearance' and acceptable for use got 1st rank. 'Maintained comfortable body posture at standing position' and "shoulder pain can be avoided' got 2nd rank. 'Good replacement to the existing tool secured 3rd rank. 'Convenient height of 'Ergo stool for harvesting and pruning "no pain and cramps in the upper region of body after performing the activity got rank 4, followed by convenient cleaning process with 5th rank and "lesser discomfort in mid back region' secured 6th rank. Similarly, Hermans *et al.* (2009) developed a cart and reported that trained, grapevine shoots reach a height of 7 feet or more. The fruit zone- range from 2 to 6 feet above ground level, again depending upon training system. These height results in many awkward hand, wrist, and back postures while working. Using a vineyard cart may improve posture by letting workers sit while carrying out different operations, such as pruning and harvesting. Although the carts are reasonably effective in preventing MSDs, they work satisfactorily; the ground is relatively flat and hard. Walters (1996) reported that the risk of hand and wrist injuries be avoided with good work habits. Reducing the hand and wrist injuries is assigning different responsibilities to workers on a given work day taking periodic breaks; working at a slower pace and using ergonomically matched tools.

Feasibility of Ergo stool for pruning and harvesting

Feasibility of Ergo stool for pruning and harvesting in terms of profitability, physical compatibility, cultural compatibility, simplicity, traibility, grip fatigue, physical stress factor, work output, tool factor and acceptability.

In profitability cent per cent of the respondents considered a ergo stool for pruning and harvesting profitable feasible technology. Most of the respondents 30 % considered the ergo stool for pruning and harvesting as feasible and while 70% respondents considered it as somewhat feasible. Technology for grape pruning, in terms of physical compatability. For cultural compatibility 80 per cent of the respondents considered it as feasible, 20 per cent respondents considered it as somewhat feasible technology. Cent per cent of the respondents considered ergo stool for pruning and harvesting as most simple feasible technology for pruning of grapes in context of simplicity and triability tool factor and accepatability.In grip fatigue 60 per cent of the respondents Ergo stool for pruning and harvesting was most feasible and for 40.0 per cent respondents it as feasible technology for reducing the grip fatigue. Cent per cent of the respondents considered ergo stool for pruning and harvesting as feasible technology for grape pruning for reducing physical stress. Majority of the respondents (90%) considered Ergo stool for pruning and harvesting as feasible, 10 per cent respondents consider it as most feasible technology, to increase the work output.

On the whole respondents, 27 percent considered Ergo stool for pruning and harvesting as most feasible 54 per cent respondents considered it as feasible, 19 per cent considered it as somewhat feasible, technology for pruning of grapes. Skelton (2007) reported that hand-picking grapes in vineyards cause a lot of awkward back postures due to the low positions of the grapes. To avoid excessive reaching an up-turned light-weight crate can be used. The handle, a half broom stick, is attached to reduce bending of the back, when moving the crate. A crate can also be used to bring the bucket or other basket to a higher level. Similarly, old orchards sledge can be used as a table to put the bucket or crate on. Stooped postures are avoided. And also a hook, used as a supporting device for carrying fruit, is also an easy-to-use solution. However, full bucket has to be lifted almost at shoulder level, which is stressful.

Feasibility index of Ergo stool for pruning and harvesting

Feasibility index was assessed in terms of profitability, physical compatibility, cultural compatibility, simplicity, triability, grip fatigue, physical stress factor, work output factor, tool factor and acceptability.

Simplicity and acceptability got first rank with feasibility index of 100%, physical stress factor and grip fatigue got 2nd rank followed by work output. Profitability, triability, culture compatibility and tool factor. The last rank was for physical compatibility.

Overall feasibility index of Ergo stool for pruning and harvesting was 82.8%, indicating that it was acceptable to the workers. Peppelman *et al.* (2006) reported that comparing pick-train, hydraulic pick-lorry by using a pick-bucket, conveyor harvester and the adjustable picking platforms (also called pluk-o-trac), the adjustable picking platforms introduced the least unhealthy working arrangements. The pick-train was the cheapest harvesting method and the workers have less static postures. The working position on the conveyor harvester can be improved, when harvesters use a crate or small ladder to pick the apples out of the top of the trees. For the picking performance, it was important that the conveyor harvester could be used in combination with another harvest method. IDEWE (External service for prevention and protection at work (non-profit) found that to transport the fruit baskets an adapted wheelbarrow was used. Lifting and carrying the baskets, were reduced. Wheel barrow was the better working height while emptying the fruit basket, the picker no longer has to bend the back so excessively. Similarly, Report on pickers prefers a basket carried on the side of the body. These allow freedom of movement during picking, but causes an asymmetrical load on the back, since the weight has to be carried on one side. During harvesting, the sleeve was fixed with a rubber band. To empty the basket, the elastic was released and the apples, pears or cherries can roll into the pallet box. APRW applied plant research Wageningen) found that the conveyor belt consists of wires on which the fruit is placed. Conveyor, fruits are transferred automatically into a pallet box. Similarly, Thiyagarajan *et al.* (2013) reported that heart rate and oxygen consumption rate of the sugarcane knives varied from 132.55 to 138 beats min-1 and 1.171 to 1.253 L min-1, respectively. The energy cost of sugarcane harvesting knives, varied from 24.45 to 26.16kJ min-1, respectively. The values of percent maximum aerobic capacity (VO2 maximum) and work pulse for sugarcane harvesting knives were much higher than that of the acceptable workload (AWL), limits of 35, the sugarcane harvesting knife (H1) ranked as I, in terms of minimum value of heat rate (132.55 beats min-1), energy cost of work (24.45 KJ min-1), acceptable workload (58.14%), over all discomfort rate (moderate discomfort) and Body part discomfort score (29.39), when compared with other three models (H2, H3 and H4) of sugarcane harvesting knives.

□□□

Chapter 6

Summary and Conclusions

Grape (Vitis vinifera) is an important fruit crop in India. Grapes are the third most widely cultivated fruit after citrus and banana. Major grape – growing states are Maharashtra, Karnataka, Andhra-Pradesh, Tamil Nadu and the north- western region covering Punjab, Haryana, Delhi, western Uttar Pradesh, Rajasthan and Madhya Pradesh (Singh, 2010). In Haryana grapes are cultivated in an area of 111.00 (000 ha) with a total production of 1235.00 (Tons) and productivity of 11.10 (tons/ha) in 2010-11 National Horticultural Board, Government of India. Haryana, is the sixth largest producers of grapes in the country with 5.7 million ton/ hectare/ year. In Haryana, it is usually cultivated from January to June end. All the result in various WMSD as in vine pruning and harvesting, the operators are open to a great deal of strain likely to trigger or accelerate the WMSDs generating mechanisms including repetitive movements combined with external forces acting on the finger, hand wrist system (Using non- powered pruning shears) and static work in the upper arm shoulder system.

Keeping in mind the risks in grape cultivation, therefore, the present study was carried out with the following objectives:

Objectives

1. To study work profile and working conditions in grapes cultivation.
2. To assess the occupational risks in grapes cultivation.
3. To develop an ergo-solution for highly risk prone activity in grapes cultivation.

Methodology

The present study was conducted in three phases:

Phase- I: Survey work - A survey was conducted to study the work profile and working conditions of workers in grapes cultivation and grape processing practices used by them.

Phase- II: Experiment work - Experiment work was done to study occupational risks of workers in grapes cultivation.

Phase- III: Ergo solution - Ergo solution was developed to overcome the hazards in the highly risks prone activity in grapes cultivation.

Phase I : Three districts of Haryana state were selected purposively in consultation with Horticultural Department of Haryana state, where grapes cultivation was being done and six grape orchards from these districts were selected randomly for the field study. The work profile of the workers was examined on the basis of type of work, time spent, rest period, body part involved, discomfort rating, pain symptoms, workload as per time and environmental condition in all the activities of grapes cultivation viz. land preparation, pruning, manuring, irrigation, plant protection, harvesting, handling and transportation.

The data were collected personally with the help of a well-structured and duly pre-tested, interview schedule. The data so collected were coded, tabulated and suitably analyzed to draw meaningful inferences.

Phase II : In phase II, a sample of 15 respondents was selected purposively from the randomly selected 2 grape orchards, out of the six grape orchards selected in phase I Respondents who were physically fit and willing to co-operate and engaged in grapes cultivation activity were selected. Physical fitness of the workers involved in grapes cultivation activity was ascertained by measuring the four physiological parameters i.e. blood pressure, body temperature, pulse rate and maximum aerobic capacity (Vo_2 Max). Environmental parameters viz. light, temperature, humidity, heat stress, suspended particulate matter were also taken occupational risk was assessed through four parameters- physical parameters, biomechanical stress, physiological load rating and time load. Physical load was assessed on the basis load rating, KIM (Key indicator method), Biomechanical stress includes grip fatigue, spinal deviation, RULA score, OWAS, repetitive strain exertion and musculo-skeletal discomfort. The physiological load was measured on the basis heart rate, oxygen consumption, energy expenditure and RPE scale. Time load was assessed on the basis of time was score.

Phase III : Ten workers of the orchard were selected purposively to obtain the feedback of developed ergo solution, i.e. Feasibility testing of Ergo solution for reducing the hazards in the highly risk prone activity in grapes cultivation.

Major findings

Phase I: Survey work

Work profile and working conditions in grapes cultivation

- Maximum respondents were in age group of above 26 years and cent per cent of the respondents were male. They all were educated upto primary and senior secondary levels and had temporary pattern of employment.
- A huge majority of respondents were doing grape farming from more than 9 years. Cent per cent of the respondents were taking rest in between the work with length of work period 4-8 hours per day. Majority of the respondents were taking rest period of 60-120 min. They were getting income (wages) weekly and working on contract basis and mode of fixing of wages of half of the respondents was based on time of work and that of other half of the respondents was on quantum of work.
- Maximum problems faced by the grape orchard workers was postural problem, injuries/cuts, workplace problem, physical problem, problem due to pesticides, problem due to work, problem due to high temperature and economic problem.
- Land preparation activity was done by during the month of December through use of *spade* in the bending posture by workers and risk prone areas of body were lower back, neck, mid back. Pruning was done during the month of January through use of secateurs in standing posture with raised arm, and risk prone areas of body were feet, lower back shoulder, and upper arms. Manuring activity was done during the month of February to April through tractor and manual spray pump, in standing posture and risk prone area were feet and lower back. Irrigation was done in the month of February to April through tube wells, in standing posture and risk prone areas were feet lower back. Harvesting activity was done during the month of May/June either manually or use of shaving blade, in standing posture with risk prone areas of feet, lower back and mid back. The workers

were packing the grapes in sitting posture without back support, and risk prone areas were lower back and mid back. Transportation activity was done in standing posture with risk prone areas in lower back, mid back and feet.

- In winter season temperature, light, humidity was below recommended value and CO_2 and suspended particles matter were above recommended value. In summer season, temperature and suspended particles matter were above recommended value and CO_2 and humidity and light were below recommended value.
- In winter season, there was no problem due to heat stress. In summer season, the heat stress in grape orchard depicts that extreme cautions was required as this heat stress results in sun-stroke, heat cramps, heat exhaustion possible with physical activity.

Phase II

- Mean height and weight of grape workers involved in grapes cultivation was 159.9 cm and 64.2 kg, respectively. Body mass Index (BMI) was observed as 21.8 Kg/m^2. Fat percentage was worked out to be 29.9%. Hence, LBM (Lean body mass) was 44.1 kg with variation of ±19.3kg. Majority of the workers (80%) were having mesomorphic (Athletic type body), which is considered as the perfect type body type.
- The grip fatigue was maximum in pruning activity for both the hands. The maximum deviation in the spinal angle was in land preparation followed by manuring, irrigation, plant protection and pruning and least grip fatigue was for harvesting and handling and transportation.
- As per the RULA, posture used in land preparation, pruning and harvesting required investigation and should be changed immediately.
- OWAS analysis depicted that posture used in pruning and harvesting was highly risky and corrective measures in posture were required immediately. The posture load factor was highest in pruning.
- As per the assessment of repetitive tasks (ART), pruning was risky activity and further investigation was required immediately.

As per strain index of the activities of grapes cultivation, pruning harvesting and handling transportation were risky activities but among these pruning scored highest score. The most repetitive activity in grapes cultivation was pruning, as it got highest repetitive strain load factor, hence risky in terms of repetitive nature of activity. On the basis of total repetitive strains score in grapes cultivation pruning was most risky activity followed by harvesting.

- Overall maximum Musculo-skeletal discomforts of workers in grapes cultivation was in harvesting activity.
- The occupational risks in terms of biomechanical stress was highest in pruning followed by harvesting. The least was in plant protection.
- The occupational risks of workers in terms of physiological load was highest for pruning followed by harvesting. The least physiological load was for plant protection.
- As per time load, pruning was highest time consuming activity.
- The overall occupational risks in grapes cultivation activities on the basis of physical load biomechanical stress, physiological load and time load was maximum in pruning and harvesting activities. Hence, ergo solution was developed for these activities.

Phase III

- On the basis of results of Phase II, pruning and harvesting of the grapes were identified as the highly risk prone activities. So, available technology used by them was modified and ergo solution was developed for these two activities. Available technology i.e. scateurs was modified by adding the materials on grip /handle. And an ergo stool for pruning and harvesting activity was developed the acceptability and feasibility of ergo stool for pruning and harvesting was assessed.
- The Acceptability of improved grape pruning secateurs highlighted that appearances and easy to work as hand grips got highest rank.
- Feasibility index of grape pruning secateurs was 93.2% indicating that it was highly feasible for them to use.

- The acceptability of ergo stool for pruning and harvesting highlighted that appearances and acceptable for use got highest rank.
- Feasibility index of ergo stool for pruning and harvesting was 82.8% indicating that it was acceptable to the user.

Suggestions and Guidelines for grapes cultivation workers

Improvement in work profile and working conditions

- Short stretching routine should be included with rest breaks after every 2 hours, to promote rest cycle so that body can recover from demands of grapes cultivation. Prolonged adoption of poor posture should be removed through correction in posture.
- Protective equipment should be used while working to reduce the incidence of injuries.

Reduction in occupational risk factor

- All the workers of grape orchard should educate on the MSI (Musculo-skeletal Injury), risk factor, trimming technique, use the safe working postures and practices during activity.
- Use the correct tool for grapes activities.
- Use of padded secateurs for different activity like pruning and harvesting should be encouraged.

Implications and recommendations of the study

The finding of the investigations brought out a number of implications for concerned users, researchers, manufactures and extension workers:

For researchers : This study will be helpful to the researchers to undertake intensive research on all the aspects about grapes cultivation workers related to their health profile, work profile and their allergic disorder, occupational risk factor, etc.

For manufactures : The findings of the study may provide useful data for those who are developing new personal protective equipment for workers to reduce the occupational risk of grapes cultivation workers.

For extension worker : This study will help the extension workers to sensitize the issues related to grapes orchards workers to develop positive attitude among them and to create interest among them to adopt safety and preventive measures.

For users : This study also help the users to make them aware about occupational risks factor encountered at work and to work and to know preventive technique to combat with these problems.

❑❑❑

Bibliography

AgSafe 1992. Occupational injuries in California agriculture, 1981-1990. UCDANR, Oakland.

AICRP 2013. Score sheet for characterization of drudgery, Deptt. of Family Resource Management, I.C. College of Home Science, CCS, Haryana Agricultural University, Hisar.

Aweto, H.A., Tella, B.A., and Johnson, O.Y. 2015. Prevalence of work-related musculo-skeletal disorders among hairdressers. ***International Journal of Occupational Medicine and Environmental Health.*** **28**(3): 545–555.

Ayub, M. and Mittal, A. 1989. Manual materials handling. London: Taylor and Francis.

Bakshi, 1997. Environment factors influencing work ergonomics in the home. Punjab Agricultural University, Ludhiana, 112p.

Bernard, C., Laurène, C., Stéphane, B., Annie, A., Jean-Claude, C. and Isabelle, N. 2011. Biomechanical and Psycho-social Work Exposures and Musculo-skeletal Symptoms among Vineyard Workers. *J Occup Health,* **53**: 297–311.

Bineford, L. 2004. Contract labour in Canda and the Unites States: A critical appreciation of Tanya basok;s tortillas and tomatoes. Transmigrant Mexican harvester in Canada. *Canadian Journal of Latin Amercian and Caribbean Studie,* **29**: 289-308.

Bonger, P.M., Kremer, M. and Luakter, J. 2002. Are psychosocial factor for symptoms and sign of the shoulder, elbows or hand and wrist. Journal of Industrial Medicine, **42**(5): 315-342.

Borah, S. 2009. Ergonomics evaluation of the rural women while performing wheat harvesting activity. *Journal Human Ecology,* **18**(4): 309-311.

Brumitt, J., Reisch, R., Krasnoselsky, K., Welch, A., Rutt, R., Garside, L.I., and Mckay, C. 2011. Self-reported Musculo-skeletal Pain in Latino Vineyard Workers. *Journal of Agromedicine,* **16**: 72–80.

Calleal, P., Zimbalatti, G., Quendler, Elsabet., Nimmerichter, A., Bachl, N., Bernardi, B., Smorto, D. and Benalia, S. 2014. Occupational illnesses related to physical strains in apple harvesting. *Annals of Agricultural and Environmental Medicine,* **21**(2): 407–411.

Carruth, A.K., Skarke, L., Moffett, B. and Prestholdt, C. 2002. Non-fatal injury experiences among women on family farms. *Clinical Nursing Research,* **11**(2): 130-148.

Casado, A.E., Soto, H.A. and Colombian, D. 2009. Repetitive movements of upper limbs in viticulture set up of annual exposure level assessment models, OCRA. *International Ergonomics Association.* 9-14.

Chapman, L.J., Newenhouse, A.C., Meyer, R.H., Taveira, A.D., Karsh, B.T., Ehlers, J.J. and Palermo, T. 2004. Evaluation of an intervention to reduce musculo-skeletal hazards among fresh market vegetable growers. *Applied Ergonomics,* **35**(1): 57-66.

Chapman,L.J., Newenhouse,A.C., Meyer,R.H., Taveira, A.D., Karsh,B.T., Ehlers,J.J. and Palermo,T. 2007. Evaluation of an intervention to reduce musculo-skeletal hazards among fresh market vegetable growers. *Applied Ergonomics.* **35**(1): 57-66.

Chatzi, L., Prokopakis, E., Tzanakis, N., Alegakis, A., Bizakis, I., Siafakas, N. and Lionis, C. 2005. Allergic rhinitis, asthama, and atopy among grape farmers in a rural population in Crete, Greece. *Occupational and Environmental Lung Diseases,* **127**: 372-378.

Dahiya, P., Singh, K., Mehta, M. and Sharma, D.K. 2013. Ergonomic Risks in Polyhouse Farming. International Conference on Ergonomics And Human Factors "Ergo 2013: Ergonomics For Rural Development, 158-167.

Dalal, P. and Acharya, M. 2010. An ergonomic assessment of beauticians' occupation and developing guidelines for work centre. Ergonomic for everyone- a collection of articles presented at the international conference on ergonomic (HWWE 2009) Calcutta, 457-462.

Donald, C. 2006. Understanding the links between agriculture and health, occupational Health Hazards of agriculture. *Vision: for Food Agriculture and the Environment,* **13:** 8-16.

Duraj, V., Miles, J.A., Meyers, J.M., Faucett, J.A., Janowitz, I.L., Tarter, M.E., Tejeda, D.G., Smith, R.H. and Weber, E.A. 2000. Harvesting aids for reducing ergonomics risk factors in wine grape hand harvesting. *Proceedings of the American Society of Agricultural Engineers Meeting.*

Estill, C.F. and Tanaka, S. 1998. Ergonomics consideration of manually harvesting maine wild strawberries. *Journal of Agricultural Safety and Health,* **4**: 43-57.

Fathallah, F.A., Meyers, J.M., Miles, J.A., Smith, R.J. and Weber, E.D. 2006. Smaller loads reduce risk of back injuries during wine grape harvest. Journal of occupational medicine. **36:** 697-717.

Fry, A.C., Ciroslan, D., Fry, M.D., Leroux, C.D. and Schilling, B.K. 2006. Anthropometric and performance Variables Discriminating Elite American Junior men weightlifters. *Journal of Strength and Conditioning Research,* **20**: 861-219.

Fulmer, S., Punnett, L., Slingerland, T. D. and Richardson, G. 2002. Ergonomic exposures in apple harvesting: preliminary observations. *American Journal of Industrial Medicine,* **2:** 3-9.

Garrow, J.S. and Webster, J. 1983. Quetelet;'s index as measures of fatness. *International Journal of Obese.* **9**: 147-153.

Ghany, A.E. H., Ismail, M.A., Nahed, K., Bialee, E. N.M. and El-Bialee, N.M. 2015. Developed hand tool to pick Prickly pear fruit. *International Journal of Advanced Research,* **3**(5): 1307-1315.

Grandejean, E.1998. Ergonomics of the Home. Taylor and Francis Ltd., London.

Guarav, M.A. and Salunkhe, S. R. 2011. A study of problem and prospects of women employees in grape farm. *Indian Stream Research Journal,* **1**: 2230-7850.

Hagberg and Sundelin, 2006. From ergo web website.*http://w. ergoweb.com/join/login.cfm.*

Hayness, S. and Williams, K. 2008. Impact of seating position on user comfort an typical performance for people with and without low back pain. *International Journal of Industrial Ergonomics,* **38**: 35-46.

Heat index by national oceanic and atmospheric administration (NOAA). National weather service office of climate and water and weather service. U.S.A.

Hermans, V., Motmans, R., O'Neill, D., Roman, D. and Lundqvist, P. 2009. Harvesting fruit. European Commission DG Employment,

Social Affairs and Equal Opportunities *http:// www.agrierg onomics.eu.*

Hildebrandt, V.H. 1995. Musculo-skeletal symptoms and workload in 12 branches of Dutch agriculture. *Ergonomics,* **38**: 2576-2587.

Hisataka, S., Miyaoa, M., Aki kondoa, T. and Yamada, S. 2007. Overhead work and shoulder-neck pain in orchard farmers harvesting pears and apples. **38**(4): 700-706.

Hokmabadi, R.A., Kavaki, K.M. and Mehdinia, M. 2012. Hairdressers by Reba: Rapid ergonomic postures of evaluation of entire body assessment. *Journal of North Khorasan University of Medical Science,* **3**(4): 49-54.

James, M.M., Janowitz, I., Tejeda, G.D., Miles, A.J., Duraj, V., Faucett, J. and Kabashima, J. 2010. Impact of selected engineering controls in reducing ergonomics risk factor associated with musculo-skeletal disorder in nursery work. *Indianapolis.*

Janowitz, I., Tejeda, D.G., Miles, J.A., Duraj, V., Fathallah, F., Meyers, J.M. and Faucett, J. 2000. Ergonomics interventions in the manual harvest of wine grapes. *Proc. of the IEA 2000/HFES 2000 Congress,* 2000.

Jeebhay, M.F., Baatjies, R. and Lopata, A.L. 2002. Occupational allergy among table grape farm workers in South Africa. *Current Allergy and Clinical Immunology,* **15**(4): 174-177.

Jeebhay, M.F., Baatjies, R. and Lopata, A.L. 2007. Occupational allergy among table grape farm workers in South Africa. *Current Allergy and Clinical Immunology,* **15**(4): 174-177.

Jeremy, F., Mike, G. and Birtles, M. 2010. Assessment of Repetitive Tasks of the upper limbs the ART tool. Health and Safety Laboratory.

Kankal and desmukh. 2014. Ergonomic Evaluation of Manual Citrus Fruit Harvester. *The International Journal Of Science and Technoledge* (ISSN 2321–919X)

Karhu, O., Kansi, P. and Kuorinka, I. 1997. Correcting working posture in industry: a practical method for analysis. *Applied. Ergonomics,* **8**: 199-201.

Kato, A.E.F.A., Fathallah, J.A., Miles, J.M., Meyers, J., Faucett, I., Janowitz, E. and Garcia, G. 2006. Ergonomic Evaluation of Wine-grape Trellis Systems Pruning Operation. *Journal of Agricultural*

Safety and Health. Journal of Agricultural Safety and Health. **15**: 174-177.

Kaur, H. and Sharma, S. 2010. Ergonomic evaluation of vegetable plucking with traditional and modern tool. *Proceedings of HWWE, 2010, Chennai, 76 pp.* (Abstract).

Kidd, P., Scharf, T. and Veazie, M. 1996. Linking stress and injury in the farming environment: A secondary analysis of qualitative data. *Health Education Quarterly,* **23**(2): 224-237.

Kittusamy, N., Mayton, A. and Ambrose, D. 2004. Self-reported musculo-skeletal symptoms among operators of farming equipment. *Proceedings of the American Industrial Hygiene Conference, Atlanta, GA.*

Kjellstrom, T. 2009. The 'Hothaps' programme for assessing climate change impacts on occupational health and productivity: an invitation to carry out field studies. Glob Health Action. **2**: 10-7.

Kroemer, K.H.E., Kroemer, H.J. and Kroemer, K.E. 1997. Engineering physiology. Van Nostrand Reinhold, New York, pp 193-228.

Kumar, S. 2000. Theories of musculo-skeletal injuries causation. *Ergonomics,* **44**: 17-47.

Kumudini, G. and Tetsuya, H. 2009. Workload and awkward posture problems among small-scale strawberry farmers in Japan. *J. Human Ergol,* **38**(81): 88-90.

Kuorinka, I. 1987. Standardized Nordic Questionnaire for analysis of musculo-skeletal system. *Applied Ergonomics,* **18:** 199-201.

Larson and Hannihen, 1995. Sewing –machine operators work and musculo-skeletal complaints. *Applied Ergonomics,* **23**(4): 268-276.

Lee, K.S., Hyo, C. K., Chae, H.S., Kim, K. Ran., Lim, D. S. 2013. Bio-mechanical Analysis of the Grapevine Cluster Thinning Task using Working Chair. *Journal of the Ergonomics Society of Korea,* **30**(3): 395-401.

Louhevaara and Suuranakki, 1992. A rapid appraisal of occupational work load modified scale of perceived exertion. *Ergonomics,* **37**(3): 485-491.

Luginbuhl, R.C., Jackson L.L., Castillo, D.N. & Loringer, K.A. (2013) Heat-related deaths among crop workers United States, *MMWR,* **57**(24): 649-653. *http://www.cdc.gov/mmwr/preview/mwrhtml/mm5724a1.*

Mayer, A. 2005. Environmental/occupational safety and health. National canter for farm worker health. *Migrant Health Issues Monograph Series*, **2:** 3-7.

Mayer, J. 2010. International Agricultural Migration and Public Health: Examining Migrant Farm Worker Health and the Public Health Implications of Agricultural Temporary Migration. Public Health Agency of Canada

Mc Atamney, L. and Corlett, E. N. 1993. RULA: A survey method for the investigation of work related upper limb disorder. *Applied Ergonomics*. **24**: 91-99.

Mclean. L., Tingly, M., Scott, R.N. and Richards. 2001. Computer terminal work and the benefits of micro breaks. *Journal of Applied Ergonomics*, **32**(3): 225-237.

Menzal, N, and cohon, S. 2007. Psyschological factors into a theoretical model for work related disorder. *Critical care Nursing Clinics of North America*, **19**(2): 143-145.

Metamney and Nigel, 1993. Agricultural work related injury and ill-health and the economic cost. *Environment Sci. and Pollut*, **6**(3): 175-182.

Meyers JM, Miles JA, Faucett J, Janowitz I, Tejeda DG, Weber E, Smith R, Garcia L. 1999.Priority risk factors for back injury in agricultural field work: vineyard ergonomics. *J Agromedicine*, **8:** 37-52.

Meyers, J. M., Miles, J. A., Faucett, J., Janowitz, I., Tejeda, D. G., Duraj, V., Kabashima, J., Smith, R. and Weber, E. 2000. High risk tasks for musculo-skeletal disorders in agricultural field work. *Proceedings of the IEA 2000/HFES Congress.*

Meyers, J.M., Miles, J.A., Faucett, J., Janowitz, I., Tejeda, G.D., Weber, E., Smith, R. and Garcia, L. 1999. Ergonomics risk factor for musculo-skeletal disorder in wine grape vineyard work. *http://ag-ergo.ucdavis.edu/paper/vineyardjmm.htm.*

Mines, R., Mullenax, N. and Saca, L. 2001. The Binational farmworker health survey: an in-depth study of agricultural worker health in Mexico and the United States: California institute for rural studies.

Miri, M.R., Hosseini, M.H. and Sharifzadeh, G.R. 2008. Evaluation of ergonomic postures of hairdressers by REBA in Birjand. *Quarterly of the Horizon Medical Sciences*, **14**(2): 39-44.

Monarca, D., Cecchini, M., Marucci, A. and Porceddu P.R. 2003. Microclimate risk for workers in greenhouses. XXX CIOSTA - CIGRV Congress, Turin, 22-24pp.

Monaria, H., Zharkova, N.S., Dobrokhotov, V. B., Burmistrova, R. A. and Lyapkalo, A. A. 2003. Hygienic characteristics of working conditions in health status of greenhouse workers. Occupational health in agriculture in the process of intensification of agricultural production Moscow, 48-51pp.

Montomoli, L., Ardissone, S., Colombini, D., Fanti, M., Ruschiani, A. and Sartorelli, L. 2010. Task analysis risk assessment and clinical Evaluation of work related musculo-skeletal disorder in Viticulture and olive culture farming. *Proceeding of International conference Ragusa SHWA Ibla Campus- Italy "Work Safety and Risk Prevention in Agro-food and Forest Systems".*

Moore, A., Wells, R. and Ranney, D. 1991. Quantifying exposure in occupational manual tasks with cumulative trauma disorder potential. *Ergonomics,* **34**: 1433–1453.

Moore, J. S. and Garg, A. 1995. The Strain Index: A proposed method to analyze jobs for risk of distal upper extremity disorders. *Journal of the American Industrial Hygiene Association,* **56**: 457-458.

Morse, T.F., Warren, N., Dhillion, C. and Diva, U. 2007. A population based survey of ergonomics risk factor in Connecticut; distribution by industry, occupation and demographics. *Journal of Connecticut Medicine,* **71**(5): 262-268.

Murali, D., Kulkarni, M. S. and Pardesi, R. S. 2004. Postural deviation of body of farm women while performing selected farm activities using traditional method and improved tools. Proceedings of HWWE, Mumbai, pp. 43. (Abstract)

National Horticultural Board, Government of India 2011. Agri-exchange.apeda.gov.in

National Institute for Occupational Safety and Health, 2001. Simple Solutions Ergonomics for Farm Workers. Report No. 1-53pp.

Newenhouse, A., Miquelon, M. and Chapman, L. 2005. Electronic pruners: faster and safer. University of Wisconsin, Madison: Healthy Farmers, Healthy Profits Project. *http://bse.wisc.edu/hfhp/tipsheets_html /pruner.htm.*

Ng, Y. G., Tamrin, S.B. M., Yusoff, I. S. M., Hashim, Z., Deros, B.M.D., Bakar, S. A. and How, V. 2015. Risk factors of musculo-skeletal disorders among oil palm fruit harvesters during early harvesting stage. *Annals of Agricultural and Environmental Medicine*, **22**(2): 286–292.

orlett, N.E and Bishop, R.P. 1976. A technique for assessing postural discomfort. *Ergonomics,* **19**: 175-182.

Parish, P., Nalavade, and Tirthankar, Ray. 2013. Comfort Analysis of Labour during Different Tea-Leaf Harvesting Practices. *International Conference on Ergonomics and Human Factors* "Ergo 2013: Ergonomics For Rural Development.

Peppelman, G., Roelofs, P.F.M.M., Schoorl, F.W. and Looije, A.A.J. 2006. Fruit Growing System Sector Fruit, Rapport, 2006-18.

Pezzi, F., Giorgio, A, Francesco, B. and Alessandro, G. 2009. Evaluation of the cutting force on vine branches in winter pruning. *J. of Ag. Eng,* **1:** 33-36.

Pheasante, S. 2010. Ergonomics Work Health, 245p.

Portier, C.J. 2012. A human health perspective on climate change: a report outlining the research needs on the human health effects of climate change. Research Triangle Park, NC: Environmental Health Perspectives/National Institute of Environmental Health Sciences. *www.niehs.nih.gov /climatereport.*

Poupart -Adam, R. Desi I., Palotas M., Vetro G., Csolle I., Nehez M., Zimanyl M., Ferke A., Huszta E., Nagymajtenyl L. 2013. Biological monitoring and health surveillance of a group of greenhouse pesticides sprayers. *Toxicol Lett,* **33:** 91-105.

Richardson, G. E., Jenkinsb, P., Fulmerc, Scott., Masona, C., Burdick, P. J.M. 2005. An ergonomic intervention to reduce back strain among apple harvest workers in New York state, *Applied Ergonomics,* **36:** 327–334.

Romano, E., Camillieri, D., Bonsignore, R., Rapisarda, V. and Schillaci, G. 2012. Analysis of stress produced during manual pruning in Vineyard: the study of anthropometric parameters. *International Conference RAGUSA SHWA 2012, September 3-6, 2012, Ragusa – Italy Safety Health and Welfare in Agriculture and in Agro-food Systems*

Roqueloure,Y., Dano, C. and Dusolier, G. 2002. Biomechanical strains on the hand–wrist system during grapevine pruning. *Int Arch Occup Environ Health,* **75**: 591–595.

Saha, P.N. 2003. Evaluation of occupational workload in a coal rolling mill in Bombay. Report No 13. SNDT, Bombay.

Sanders and McCormic.1993. Human Factors in Engineering and Design. Seventh Ed, McGraw Hill, Inc., New York.

Shikamany, S.D. 2003. Grape Cultivation in India. FAO.

Shikhamany, S D. 2001. Grape production in India. Grape Production in the Asia-Pacific Region, Minas K Papademetriou and Frank J Dent (Eds). FAO Regional Office for Asia and the Pacific, Thailand, pp. 28-38.

Singh, A. (2010). Recent Initiative in Horticulture, Westville Publishing House, Ist Edition, pp. 2-4.

Singh,S.P. and Gite, L.P. 2007. Ergonomics evaluation of hand operated paddy winnower by women worker. *Journal of Agricultural Engineering*, **44**: 67-71.

Skelton, S. 2007. Viticulture: an introduction to commercial grape growing for wine production, London, 238p.

Smith, A.B., Shimomur, Y., Iwanag, K. and Katsuura, T. 2006. The influence of different lower seat height on the muscular stress during squatting for a ground level job. *Journal of Human Ergology*, **36**: 25-33.

Sriwarno, B. A., Shimomura, Y., Iwanaga, K. and Katsuura, T. 2006. The evaluation of the effect of stool height alteration on workload of squatting postures performed by Indonesians with different body mass index (BMI). *J. Human Ergol*, **35**: 31-39.

Steinberg, U., Liebers, A., Klubmann, H.J., Grbhardt, M.A., Behrendt, R.S. and Latza, U. 2012. Key indicator method for assessing manual workload during manual handling operations. 1st editions. Published by the federal institute for occupational safety and health, pp. 88261-7221.

Stueland, D.T., Lee, B.C., Nordstrom, D.L., Layde, P.M., Wittman, L.M. and Gunderson, P.D. 1997. Case-control study of agricultural injuries to women in central Wisconsin. *Women and Health*, **25**(4): 91-103.

Takala. 2008. European Agency for Safety and Health at Work. Work-related musculoskeletal disorders: Prevention report. Luxembourg: European Agency for Safety and Health at Work.

Thiyagarajan, R., Kathirvel, K. and Jayashree, G.C. 2013. Ergonomic intervention in sugarcane harvesting knives. *African Journal of Agricultural Research,* **8**(6): 574-581.

Van, D., Molen, H.F., Sluiter, J.K., Hulshof, C.T., Vink, P. and Frings, M.H. 2005. Implmentation of participatory ergonomics intervention in construction companies. *Scandinavian Journal of Work Environment Health,* **31**: 191-204.

Varghese, M.A., Saha, P.N. and Atrey, N. 1994. A rapid appraisal of occupational work load from a modified scale of perceived exertion. *Ergonomics,* **37**: 485-491.

Villarejo, D., Lighthall, D., Williams, D., Souter, A., Mines, R., Bade, B., Samuels, S. and McCurdy, S. 2000. Suffering in silence: A report on the health of california's agricultural workers. Woodland Hills, CA: *The California Endowment.*

Vinayak, R.N., Singh, P. and chahal, V.P. 2015. Attitude of farmers towards grapes cultivation and export. *Indian Journal of Agricultural Sciences,* **85**(4): 592–5.

Wakula, J., Beckmann, T., Hett, M. and Landau, K. 2000. Ergonomics analysis of grapevine pruning and wine harvesting to define work and hand tools design requirements. *Occupational Ergonomics,* **2**: 151-161.

Walters, J.E. 1996. Gardening may trigger carpal tunnel syndrome. The Associated Press. *http://www.highbeam.com/doc/1P2-22847448.html.*

Williams, R.M., Schmuck, G., Allwood, S., Sanchiz, M., Shea, R. and Wark, G. 2007. Psychometric evaluation of health related work outcomes measures for musculo-skeletal disorder: *A Systematic review Journal of Occupational Rehabilitation,* **6**(1): 22-29.

Work Safe BC. 2009. Facts: Agriculture in BC. *(http://www.worksafebc.com/news room/news_releases/assets/nr_07_03_12/noi_su mmaries.pdf.*

Youakim, S. 2006. Occupational health risks of wine industry workers. *Bc medical Journal,* **48**(8): 386-391.

Yu, I.T. and Wong, T.W. 1996. Musculo-skeletal problems among VDU workers in a Hong Kong Bank. *Occupational Medicine,* **46**(4): 275-526.

□□□

Annexures

Annexure- I

1. Name of the respondents
2. Age (years)
3. Gender
4. Education qualification
    a) Read write
    b) Primary
    c) Sen. Secondary
    d) College
    e) Diploma
5. Pattern of employment
    a) Permananent
    b) Temporary
    c) Daily labour
6. Source of income
    a) Agriculture
    b) Private job
7. Family type
    a) Joint
    b) Nuclear
8. Family size
    a) 1-3 members (small)
    b) 4-5 members(medium)
    c) Above 5 (large)
9. Marital staus
    a) Married
    b) Single
    c) Widowed

10. Income
    a) Low (Upto ₹70,000) annum)
    b) Medium (70,001-140,000) per annum
    c) High (More than 140,000) per annum
11. Caste

    Low : Chamar, Bhangi, Dom, Dhobi, Badi, Jimar

    Medium : Lohar, Kumhar, Dhayal, Baniya, Sonar, Ahir , Saini, Arora

    High : Brahmin, Jat, Rajput, Bishoni
12. Land holding
    a) Landless b) less than 1 acre
    c) Marginal(5-10) d) Large (10 acres)
13. Possesson of animal
    a) Cow b) Buffalo
14. Type of land for grape cultivation
    a) Irrigated b) Dry
    c) Own/lease
15. Type of grape orchard farming
    a) Own land b) Lease
16. Working hours of workers in grape orchards
    a) Years of grape farming
    b) Rest in between the work yes/no
    c) Length of work period
    d) Length of rest period
17. Mode of payment
    a) **Wage are paid :** Daily/ Weekly/ Monthly
    b) **Mode of fixing wage** : Time of work/ Quantum of work/ Any other (specify)

18. Level of tiredness and pain faced by grapes orchards workers

Level of tiredness and pain	Hisar(n=18)	Sirsa(n=8)	Fatehabad(n=6)	Total(n=32)
Feeling tired while doing this activity				
Yes				
After how many hours you start feeling tired				
1-2hrs				
2-3hrs				
3-4hrs				
More than 4hrs				
Feeling body pain and discomfort				
Sometimes				
Always				
Frequently				
When you feel pain				
While doing work				
After work				
Treatment taken for pain and discomfort				
Massage				
Medicine				
Rest				

19. Level of problems faced by grape orchard workers

Problems	Hisar (n=18)	Sirsa (n=8)	Fatehabad (n=6)
Physical problems			
Back-ache due to prolong working Asthama Skin problems Pain in legs Pain in hand Numbness in body Swelling in hands Tingling in hands			
Work place problems			
Lack of cleanliness Lack of safety measures			
Problems due to high tempearture			
Burn skin Eye strain Vomiting Headache			
Problems related to work			
Hard physical work Excretion due to over work Rude behaviour of owner Rude behaviour of co-worker Abusive language			
Problem due to Pesticides			
Sneezing Breathlessness Eye irritation			
Economics problems			
Delay in payment wages Low wages			
Postural problems			
Pain due to stiffness of joints Pain in toes due to prolong standing Unable to stand properly			
Injuries/cuts			
Cut Fracture Wound Fall Insects biting			

20. Area of grape orchard

District	Length(m)	Width(m)	Area of grape orchard(m^2)
Hisar			
Sirsa			
Fatehabad			

21. Varieties of grapes cultivated

Varieties	Hisar	Sirsa	Fatehabad	Total
Anab-e-shahi				
Perletee				
Thomson seedless				
Cultivation in grapes orchards				
Raising				
Training system				
Telephone system				

22. Training profile of the owner in orchards

	Hisar	Sirsa	Fatehabad	Total
Have you taken any training?				
Yes				
Type of training				
Workshop				
Lecture				
From whom learnt about growing grapes cultivation				
Family				
Friend				
Self-experience				

23. Knowledge and practices of grape orchard owner about grape processing

Grapes harvested ?t for sale	Hisar	Sirsa	Fatehabad	Total
Upto 1 day				
Upto 2 days				
Upto 3 days				
Knowledge about grape processing				
Yes				
Use the grapes unsold				
Raisins				
Wine				
Chutney				
Do not know				
Discard				
Use of seedless grape for making for grape product				
Yes				
No				
Grape product made through processing of grapes				
Raisins				
Jam				
Juices				
Pickle				
Wine				
Use the grapes leaves for chutney				
Yes				
Use any other part of grape plant for making products				
No				
Grape processing unit near orchard				
No				
Yes				

23. Posture used and risk prone areas of body in activities performed by the workers in grapes cultivation

Activities	No. of workers	Work schedule	Period of activity	Posture adopted	Risk prone areas of body	Posture	Technology
Land preparation							
Planting							
Training							
Pruning							
Manuring							
Irrigation							
Trimming							
Picking							
Packing							
Transportation							

24. Environmental parameters in Grapes orchards

Parameters	Morning			Mean	Noon			Mean	Evening			Mean
	R_1	R_2	R_3		R_1	R_2	R_3		R_1	R_2	R_3	
Tem												
Hum												
Light												
SPM												
CO_2												
Heat stress												

HEAT INDEX

	ENVIRONMENTAL TEMPERATURE (F°)										
	70°	75°	80°	85°	90°	95°	100°	105°	110°	115°	120°
Relative Humidity	**Apparent Temperature***										
0%	64°	69°	73°	78°	83°	87°	91°	95°	99°	103°	107°
10%	65°	70°	75°	80°	85°	90°	95°	100°	105°	111°	116°
20%	66°	72°	77°	82°	87°	93°	99°	105°	112°	120°	130°
30%	67°	73°	78°	84°	90°	96°	104°	113°	123°	135°	148°
40%	68°	74°	79°	86°	93°	101°	110°	123°	137°	151°	
50%	69°	75°	81°	88°	96°	107°	120°	135°	150°		
60%	70°	76°	82°	90°	100°	114°	132°	149°			
70%	70°	77°	85°	93°	106°	124°	144°				
80%	71°	78°	86°	97°	113°	136°					
90%	71°	79°	88°	102°	122°						
100%	72°	80°	91°	108°							

*Combined index of heat and humidity...what it "feels like" to the body.

APPARENT TEMPERATURE	HEAT STRESS RISK WITH PHYSICAL ACTIVITY AND/OR PROLONGED EXPOSURE
90° - 105°	Heat cramps or heat exhaustion possible
105° - 130°	Heat cramps or heat exhaustion likely, Heatstroke possible
130° and up	Heatstroke highly likely

Annexure- II

Occupational risks of workers in grapes cultivation Physical parameters assessment:

Physical parameters	Before activity			After activity		
Height						
Weight						
Body composition						
VO_2max(ml/kg* min						
Body type						
Body temperature						
Blood pressure						
Pulse rate						
Pulse pressure						

Workers involved in grapes cultivation as per load rating

S.No.	Item	Score/ Rating	
1.	Load rating	Very heavy	5
		Heavy	4
		Moderately heavy	3
		Low	2
		Very low	1

Work load assessment

Workload assessment was done through Key Indicator Method (KIM)

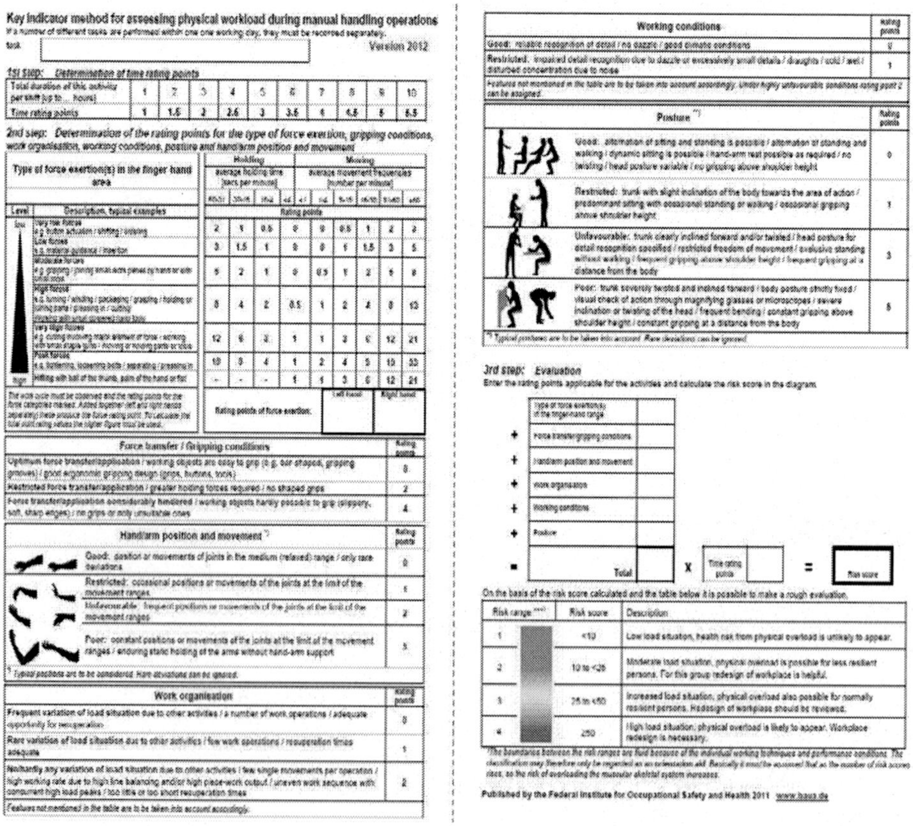

Key indicator method for assessing physical workload during manual handling operations

If a number of different tasks are performed within one one working day, they must be recorded separately.

task: Version 2012

1st step: Determination of time rating points

Total duration of this activity per shift [up to ... hours]	1	2	3	4	5	6	7	8	9	10
Time rating points	1	1.5	2	2.5	3	3.5	4	4.5	5	5.5

2nd step: Determination of the rating points for the type of force exertion, gripping conditions, work organisation, working conditions, posture and hand/arm position and movement

Type of force exertion(s) in the finger hand area — Holding: average holding time [secs per minute]; Moving: average movement frequencies [number per minute]

Level	Description, typical examples	Rating points								
low	Very low forces, e.g. button actuation / shifting / ordering	2	1	0.5	0	0	0.5	1	2	3
	Low forces, e.g. material guidance / insertion	3	1.5	1	0	0	1	1.5	3	5
	Moderate forces, e.g. gripping / joining small work pieces by hand or with small tools	5	2	1	0	0.5	1	2	4	8
	High forces, e.g. turning / winding / packaging / grasping / holding or joining parts / pressing in / cutting / working with small powered hand tools	8	4	2	0.5	1	2	4	8	13
	Very high forces, e.g. cutting involving major element of force / working with small staple guns / moving or holding parts or tools	12	6	3	1	1	3	6	12	21
	Peak forces, e.g. tightening, loosening bolts / separating / pressing in	18	9	4	1	2	4	5	10	33
high	Hitting with ball of the thumb, palm of the hand or fist	-	-	-	1	1	3	6	12	21

The work cycle must be observed and the rating points for the force categories marked. Added together (left and right hands separately) these produce the force rating point. To calculate the total point rating values the higher figure must be used.

Rating points of force exertion: Left hand / Right hand

Force transfer / Gripping conditions	Rating points
Optimum force transfer/application / working objects are easy to grip (e.g. bar shaped, grasping grooves) / good ergonomic gripping design (grips, buttons, tools)	0
Restricted force transfer/application / greater holding forces required / no shaped grips	2
Force transfer/application considerably hindered / working objects hardly possible to grip (slippery, soft, sharp edges) / no grips or only unsuitable ones	4

Hand/arm position and movement *)	Rating points
Good: position or movements of joints in the medium (relaxed) range / only rare deviations	0
Restricted: occasional positions or movements of the joints at the limit of the movement ranges	1
Unfavourable: frequent positions or movements of the joints at the limit of the movement ranges	2
Poor: constant positions or movements of the joints at the limit of the movement ranges / enduring static holding of the arms without hand-arm support	3

*) Typical positions are to be considered. Rare deviations can be ignored.

Work organisation	Rating points
Frequent variation of load situation due to other activities / a number of work operations / adequate opportunity for recuperation	0
Rare variation of load situation due to other activities / few work operations / recuperation times adequate	1
No/hardly any variation of load situation due to other activities / few single movements per operation / high working rate due to high line balancing and/or high piece-work output / uneven work sequence with concurrent high load peaks / too little or too short recuperation times	2
Features not mentioned in the table are to be taken into account accordingly.	

Working conditions	Rating points
Good: reliable recognition of detail / no dazzle / good climatic conditions	0
Restricted: impaired detail recognition due to dazzle or excessively small details / draughts / cold / wet / disturbed concentration due to noise	1
Features not mentioned in the table are to be taken into account accordingly. Under highly unfavourable conditions rating point 2 can be assigned.	

Posture **)	Rating points
Good: alternation of sitting and standing is possible / alternation of standing and walking / dynamic sitting is possible / hand-arm rest possible as required / no twisting / head posture variable / no gripping above shoulder height	0
Restricted: trunk with slight inclination of the body towards the area of action / predominant sitting with occasional standing or walking / occasional gripping above shoulder height	1
Unfavourable: trunk clearly inclined forward and/or twisted / head posture for detail recognition specified / restricted freedom of movement / exclusive standing without walking / frequent gripping above shoulder height / frequent gripping at a distance from the body	3
Poor: trunk severely twisted and inclined forward / body posture strictly fixed / visual check of action through magnifying glasses or microscopes / severe inclination or twisting of the head / frequent bending / constant gripping above shoulder height / constant gripping at a distance from the body	5

**) Typical postures are to be taken into account. Rare deviations can be ignored.

3rd step: Evaluation

Enter the rating points applicable for the activities and calculate the risk score in the diagram.

Type of force exertion(s) in the finger/hand range
+ Force transfer/gripping conditions
+ Hand/arm position and movement
+ Work organisation
+ Working conditions
+ Posture
= Total × Time rating points = Risk score

On the basis of the risk score calculated and the table below it is possible to make a rough evaluation.

Risk range ***)	Risk score	Description
1	<10	Low load situation, health risk from physical overload is unlikely to appear.
2	10 to <25	Moderate load situation, physical overload is possible for less resilient persons. For this group redesign of workplace is helpful.
3	25 to <50	Increased load situation, physical overload also possible for normally resilient persons. Redesign of workplace should be reviewed.
4	≥50	High load situation, physical overload is likely to appear. Workplace redesign is necessary.

***) The boundaries between the risk ranges are fluid because of the individual working techniques and performance conditions. The classification may therefore only be regarded as an orientation aid. Basically it must be assumed that as the number of risk scores rises, so the risk of overloading the muscular skeletal system increases.

Published by the Federal Institute for Occupational Safety and Health 2011 www.baua.de

Physical load of workers in grapes cultivation

Farm Activity	Physical load / acre				
	Weight of the load (kgs)	Distance carried (kms)	Height lifted (mts)	Physical load rating[1] (score)	Physical load factor (Total score)
Land preparation					
Removing of stalks and stubbles and unwanted plants					
Ploughing					
Pruning					
Cutting of undesirable vines					
Manuring					
Transportation of manure					
Mixing of manure					
Spreading of manure					
Irrigation					
Preparation of irrigation channels					
Plant protection					
Spraying					
Topping					
Harvesting					
Fruit picking					
Gathering and heaping					
Packaging in polythene					
Trimming					
Handling and transportaion					
Loading of the product					

Occupational risks of grape orchards workers in terms of physical load

Activities	Physical load			Total physical load	Rank
	Load rating	KIM	Physical load factor		
Land preparation					
Pruning					
Manuring					
Irrigation					
Plant protection					
Harvesting					
Handling and transportation					

Occupational risk in grapes cultivation as per Biomechanical Stress
Grip strength

Grip strength	Before activity			After activity		
Left hand						
Right hand						

Postural analysis
Angle of deviation of the spine while working in before and after activity

		Angle at normal position	**Angle at working**	**Increase over base**
Spinal angle (?exi curve)	Before activity After activity			
Back (inclinometer)	Before activity After activity			

RULA scores while grape cultivation

Posture	**Score A (Wrist and arm score +force+ Muscle used)**	**Score B (Neck+trunk and leg score+ force+ muscle used)**	**Final score**	**Action category**

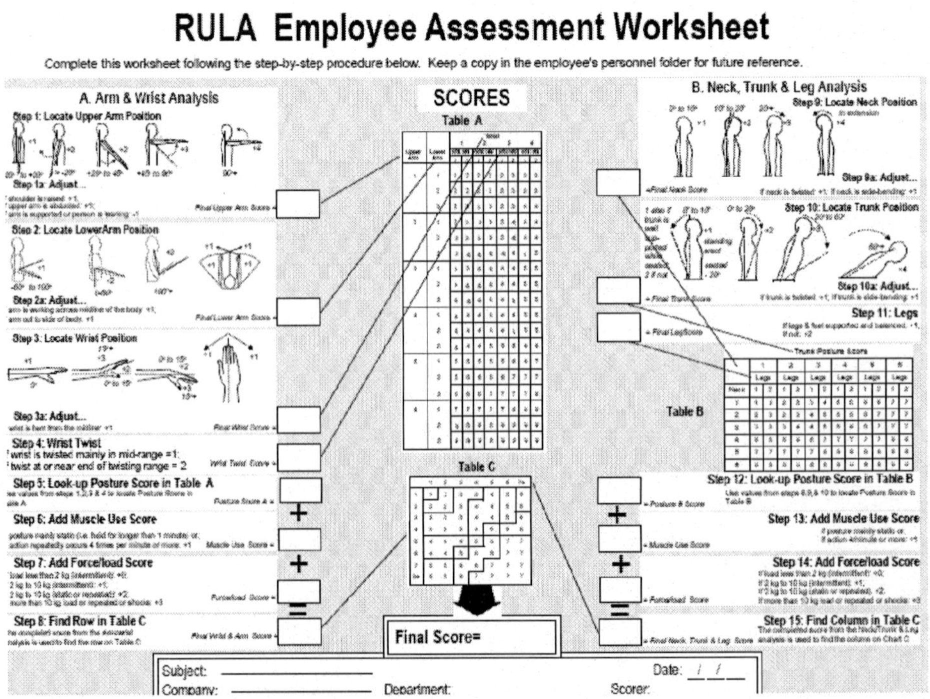

OWAS Scores while grapes cultivation

Posture	back	Upper limb	load	Final score	Action category

BACK

1 = straight
2 = bent forward, backward
3 = twisted or bent sideways
4 = bent and twisted or bent forward and sideways

ARMS

1 = both arms are below shoulder level
2 = one arm is at or above shoulder level
3 = both arms are at or above shoulder level

Posture load score of workers in grapes cultivation

Farm Activity	Nature of posture	Body part involved													Posture load factor (Total Score)
		Head	Neck	Shoulder	Upper arms	Lower arms	?ngers	hands	Waist	Hips	Upper legs	Lower legs	Upper back	Lower back	
Land preparation															
Removing of stalks and stubbles and unwanted plants															
Ploughing by country plough															
Pruning															
Cutting of undesirable vines															
Manuring															
Transportations of manure															
Mixing of manure															
Spreading of manure															
Preparation of irrigation															
Plant protection															
Spraying															
topping															
Harvesting															
Fruit picking															
Gathering and heaping															
Packaging in polythene															
Trimming															
Handling and transportation															
Loading of the product															

ART (tools) assessment of repetitive tasks of the upper limb) in grapes cultivation activities

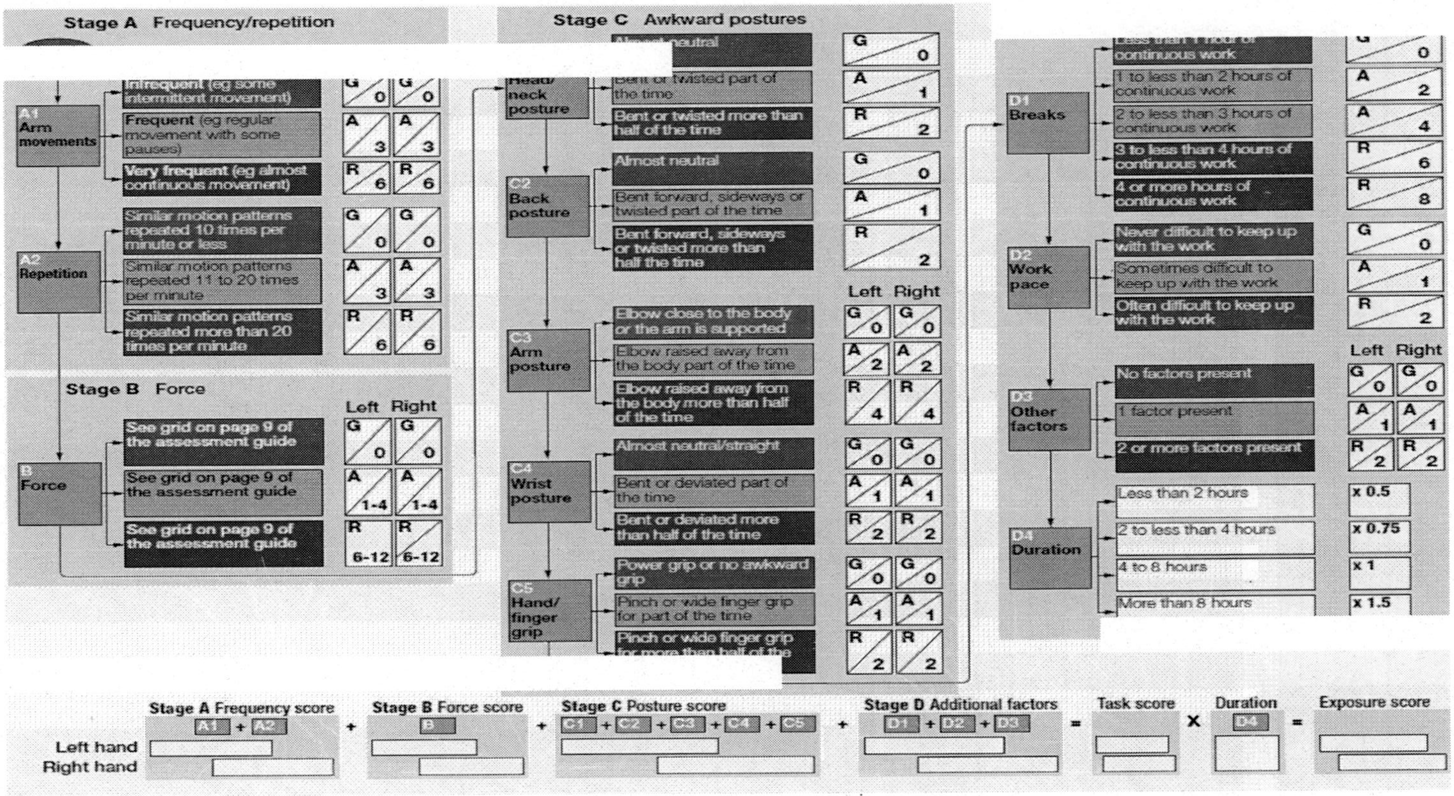

Strain Index Scoring Sheet

Date:		Task:	
Company:		Supervisor:	
Dept:		Evaluator:	

Risk Factor	Rating Criterion	Observation			Multiplier	Left	Right
Intensity of Exertion (Borg Scale - BS)	Light	Barely noticeable or relaxed effort (BS: 0-2)			1		
	Somewhat Hard	Noticeable or definite effort (BS: 3)			3		
	Hard	Obvious effort; Unchanged facial expression (BS: 4-5)			6		
	Very Hard	Substantial effort; Changes expression (BS: 6-7)			9		
	Near Maximal	Uses shoulder or trunk for force (BS: 8-10)			13		
Duration of Exertion (% of Cycle)	< 10%	**Calculated Duration of Exertion** (from inputs below)			0.5		
	10-29%	**User Inputs**	**Left**	**Right**	1.0		
	30-49%	Total observation time (sec.)			1.5		
	50-79%	Single exertion time (sec.)			2.0		
	≥ 80%	Number of exertions during observation time			3.0		
	Calculated Duration of Exertion (%)						
Efforts Per Minute	< 4	**Calculated Efforts Per Minute** (from inputs above)			0.5		
	4 - 8		**Left**	**Right**	1.0		
	9 - 14				1.5		
	15 - 19				2.0		
	≥ 20				3.0		
Hand/Wrist Posture	Very Good	Perfectly Neutral			1.0		
	Good	Near Neutral			1.0		
	Fair	Non-Neutral			1.5		
	Bad	Marked Deviation			2.0		
	Very Bad	Near Extreme			3.0		
Speed of Work	Very Slow	Extremely relaxed pace			1.0		
	Slow	Taking one's own time			1.0		
	Fair	Normal speed of motion			1.0		
	Fast	Rushed, but able to keep up			1.5		
	Very Fast	Rushed and barely/unable to keep up			2.0		
Duration of Task Per Day (hours)	<1				0.25		
	1 < 2				0.50		
	2 < 4				0.75		
	4 ≤ 8				1.00		
	> 8				1.50		
Results Key		SI ≤ 3	Job is probably safe				
		3 < SI < 7	Job may place individual at increased risk for distal upper extremity disorders				
		7 ≤ SI	Job is probably hazardous				

Repetitive strain load score of workers in grapes cultivation

Farm activity	Nature	Repetitive strain rating[3]	Repetitive strain load score	Mean repetitive strain load score
Land preparation				
Removing of stalks and stubbles and unwanted plants				
Ploughing				
Pruning				
Cutting of undesirable vines				
Manuring				
Transporation of manure				
Mixing of manure				
Spreading of manure				
Irrigation				
Preparing channel for irrigation				
Plant protection				
Covering with net				
Spraying				
Topping				
Harvesting				
Fruit picking				
Gathering and heaping				
Packaging in polythene				
Trimming				
Handling and transportaion				
Loading of the product				

Total Repetitive strains (exertion) of workers in grapes cultivation

Activities	Repetitive strains			Total repetitive strain (exertion) score	Rank
	ART score	Strain index score	Repetitive strain load score		
Land preparation					
Pruning					
Manuring					
Irrigation					
Plant protection					
Harvesting					
Handling and transportation					

Musculoskletal discomfort in which body part you feel the pain

Body part	Human body map (5 point scale) 1 2 3 4 5	VAD scale (10 point scale) 1 2 3 4 5 6 7 8 9 10
Eye		
neck		
shoulder		
Upper arm		
elbows		
Lower arms		
Wrist/hands		
buttocks		
knees		
Calf muscles		
Ankle/feet		

❒ 1= very light, 2= light, 3= moderately heavy, 4= heavy, 5= very heavy

Nordic Scale

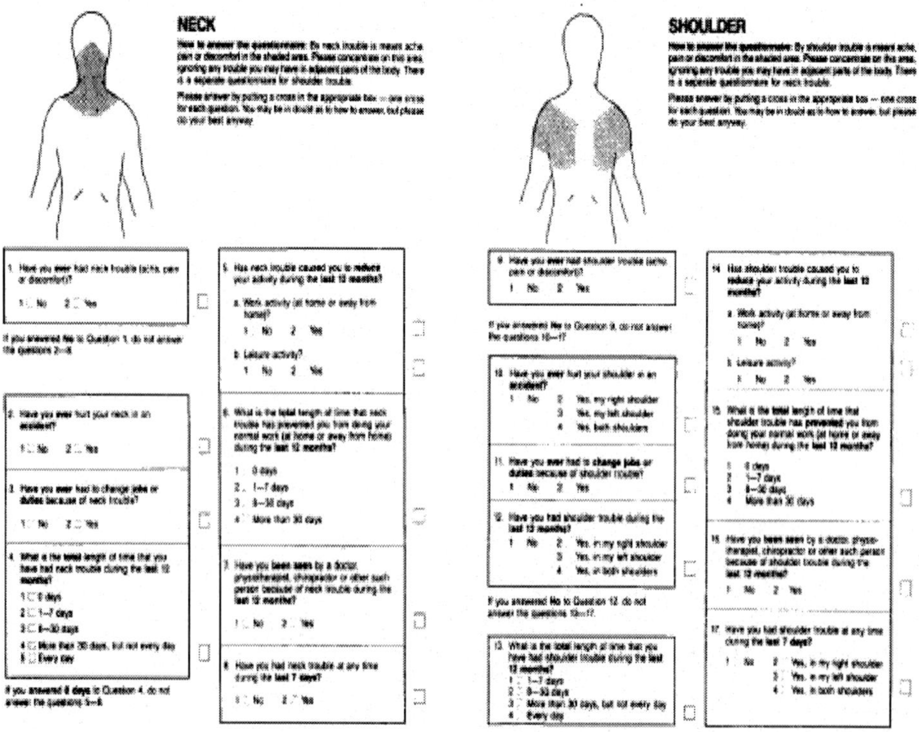

Heart Rate

Respondent	Resting HR	Working HR	Recovery HR	Duration of activity
1				
2				
3				
4				
5				
6				
7				
8				
9				
10				
11				
12				
13				
14				
15				

RPE Scores while grapes cultivation

Activity	**Discomfort Score**
Land preparation	
Pruning	
Manuring	
Irrigation	
Plant protection	
Harvesting	
Handling and transportation	

Physiological load of workers in grapes cultivation activities

Farm activity	Physiological load rating[4]	Physiological load factor	Mean physiological load factor
Land preparation			
Removing of stalks and stubbles unwanted plants			
Ploughing			
Pruning			
Cutting of undesirable vines			
Manuring			
Transportation of manure			
Mixing of manure			
Spreading of manure			
Irrigation			
Preparation of irrigation			
Plant protection			
Covering with net			
Spraying			
Topping			
Harvesting			
Fruit picking			
Gathering and heaping			
Packaging in polythene			
Trimming			
Handling and transportation			
Loading of the product			

Time load factor of workers in grapes cultivation

Farm Activity	Duration / Time						Mean Time load factor
	Hours/ day	No. of days	No. of man days/ season[5]	No. of labour employed	Work-load as per time[6]	Time Load Factor (Total score)	
Land preparation							
Removing of stalks and stubbles unwanted plants							
Ploughing							
Pruning							
Cutting of undesirable vines							
Manuring							
Transporation of manure							
Mixing of manure							
Spreading of manure							
Irrigation							
Preparation of irrigation channels							
Plant protection							
Covering with net							
Spraying							
Topping							
Harvesting							
Fruit picking							
Gathering and heaping							
Packaging in polythene							
Trimming							
Handling and transportation							
Loading of the product							

Score sheet

S.No.	Item	Score/ rating	
1.	Load rating	Very heavy	5
		Heavy	4
		Moderately heavy	3
		Low	2
		Very low	1
2.	Nature of posture	Standing	1
		Bending	2
		Twisted trunk	3
		Sitting	4
		Squatting	5
		Kneeling	6
3	Body parts involved	Neck	1
		Shoulder	2
		Upper arm	3
		Elbow	4
		Lower arm	5
		Wrist	6
		Hands	7
		Chest abdomen	8
		Upper back	9
		Lower back	10
		Buttocks	11
		Hips	12
		Upper legs	13
		Knees	14
		Lower legs	15
		Ankles	16
		Feet / nails	17
		Eye irritation	18
4.	Discomfort rating	Very severe	5
		Severe	4
		Moderate	3
		Mild	2
		Very mild	1
5.	Repetitive strain nature	Continuous	1
		Cyclic	2
6.	Repetitive strain rating	Very exhausted	5
		Exhausted	4
		Moderately exhausted	3
		Mildly exhausted	2
		Comfortable	1
7.	Physiological workload rating	Very light	1
		Light	2
		Moderately heavy	3
		Heavy	4
		Very heavy	5
8.	Pain symptoms	Itching	1
		Burning	2
		Numbness	3
		Stiffness	4
		Swelling	5
		Pain	6
9.	Body pain rating	Very painful	5
		Painful	4
		Moderate	3
		Mild	2

Overall occupational risk of workers in grapes cultivation

Activities	Physical load	Biomechanical stress	Physiological load	Time load	Overall occupational risk
Land preparation					
pruning					
Manuring					
Irrigation					
Plant protection					
Harvesting					
Handling and transportation					

Annexure- III

Acceptability of Improved Grape pruning and Harvesting secateurs

Name :

Designation :

Place of work :

Tick the appropriate option :

S. No.	Parameter	Very much	Somewhat	Not at all
1	Liked the appearance			
2	Comfortable size and shape of the grip			
3	Easy to work, as hand grips were convenient			
4	Soft and suitable grip			
5	Minimum twisting of trunk while doing the activity			
6	Maintained comfortable body posture at standing position			
7	Restrained movements of hands			
8	Increased coverage area under harvesting and pruning			
9	Good replacement to the existing tool and technique			
10	Pruning can be done in proper way			
11	Multipurpose in usage			
12	Acceptability for use			

Acceptability of Ergo stool for pruning and harvesting

Sr. No.	Parameter	Very much	Somewhat	Not at all
1	Liked the appearance			
2	Convenient cleaning process			
3	Convenient height of Ergo stool for pruning and harvesting			
4	Maintained comfortable body posture at standing position			
5	Shoulder pain can be avoided			
6	Lesser discomfort in mid back region			
7	No pain and cramps in the upper region of body after performing the activity			
8	Good replacement to the existing tool			
9	Acceptability for use			

Perceived feasibility of the improved grape pruning secateurs profitability.

In your opinion how much profitable is the improved grape pruning secateurs in terms of saving time and energy	Most feasible (3)	feasible (2)	Somewhat feasible (1)

Physical compatibility

To what extend do you feel improved grape pruning scateurs being physical compataible in terms of risk, practability, local resource utilization	Most feasible (3)	Feasible (2)	Somewhat feasible (1)

Cultural compatibility

Is improved grape pruning secateurs culturally compatibility based on past experience and existing value	Most feasible (3)	Feasible (2)	Somewhat feasible (1)

Simplicity

Give your opinion as to what extend improved grape pruning scateurs is easy or difficult to use	Very feasible (3)	Feasible (2)	Somewhat feasible (1)

Triability

Give your opinion as to what extend improved grape pruning scateurs is easy or difficult to try	Very feasible (3)	Feasible (2)	Somewhat feasible (1)

Grip fatigue

	Very feasible (3)	**Feasible (2)**	**Somewhat feasible (1)**
able to work easily as the handgrips were positioned conveniently			

Physical stress factor

	Very feasible (3)	**Feasible (2)**	**Somewhat feasible (1)**
The activity is light enough when I use the improved grape pruning scateurs/ergo stool for pruning and harvesting			

Work output factors

	Very feasible (3)	**Feasible (2)**	**Somewhat feasible (1)**
The improved grape pruning scateurs/ ergo stool for pruning and harvesting is peffective as per time cost			

Tool factor

	Very feasible (3)	**Feasible (2)**	**Somewhat feasible (1)**
Size of improved grape pruning scateurs is not comfortable to fasten as well to my body			

Acceptability

	Very feasible (3)	**Feasible (2)**	**Somewhat feasible (1)**
Improved grape pruning scateurs/ergo stool for pruning and harvesting with modification is good replacement to existing tool technique.			

Annexure-IV

	Hisar District		
Sr.no	**Name**	**Address**	**Phone number**
1	Atar Singh (owner)	Village-Aryanagar, (Hisar) Atar Singh danghi Gangua road, nearby Electricity board	9991751906
	Satbir Singh	Village-Aryanagar, (Hisar) Satbir Singh danghi Gangua road , nearby Electricity board	9992159341
2	Zile Singh	Village-Aryanagar, (Hisar) Zile Singh danghi Gangua road , nearby Electricity board	-
3	Sarup Singh	Village-Aryanagar, (Hisar) Sarup Singh danghi Gangua road , nearby Electricity board	-
4	Birj Singh	Village-Aryanagar, (Hisar) Birij Singh danghi Gangua road, nearby Electricity board	-
5	Kapoor Singh	Village-Aryanagar, (Hisar) Kapoor Singh danghi Gangua road, nearby Electricity board	-
6	Lilu Ram	Village-Aryanagar, (Hisar) Lilu Singh danghi Gangua road, nearby Electricity board	-
7	Chatur Singh	Village-Aryanagar, (Hisar) Chatur Singh danghi Gangua road, nearby Electricity board	-
8	Krishn Singh	Village-Aryanagar, (Hisar) Krishn Singh danghi Gangua road, nearby Electric ity board	-
9	Satlal Singh	Village-Aryanagar, (Hisar) Satlal Singh danghi Gangua road, nearby Electricity board	
11	Jitendra Singh	Village-Aryanagar, (Hisar) Jitendra Singh danghi Gangua road, nearby Electricity board	-
12	Ram Lal Yadav	Village-Aryanagar, (Hisar) Jitendra Singh danghi Gangua road, nearby Electricity board	-
13	Vinod Kumar	Village-Aryanagar, (Hisar) Jitendra Singh danghi Gangua road, nearby Electricity board	-
14	Manohar Lal	Village-Aryanagar, (Hisar) Jitendra Singh danghi Gangua road, nearby Electricity board	-
15	Virendra Kumar	Village-Aryanagar, (Hisar) Jitendra Singh danghi Gangua road, nearby Electricity board	-
16	Jamo Ram	Village-Aryanagar, (Hisar) Jitendra Singh danghi Gangua road, nearby Electricity board	-
17	Dinanath	Village-Aryanagar, (Hisar) Jitendra Singh danghi Gangua road, nearby Electricity board	-
18	Ranjeet Singh	Village-Aryanagar, (Hisar) Jitendra Singh danghi Gangua road, nearby Electricity board	-

List of grape orchards owner and workers in three Districts

	Sirsa District		
Sr.no	**Name**	**Address**	**Phone number**
1	Arvind Bansal (Owner)	Near Ding road, District Sirsa	9729047422
2	Data Ram	Near Ding road, District Sirsa	9813112679
3	Banvir Singh	Near Ding road, District Sirsa	-
4	Hargovind	Near Ding road, District Sirsa	-
5	Murarilal	Near Ding road, District Sirsa	-
6	Tarsim Lal	Near Ding road, District Sirsa	-
7	Rakesh	Near Ding road, District Sirsa	-
8	Aatma Ram	Near Ding road, District Sirsa	-

	Fathebad District		
Sr.no	**Name**	**Address**	**Phone number**
1	Subhash	Village-Dolu,Raitya road	9416714486
2	Mahavir Singh	Village-Dolu,Raitya road	9416163100
3	Vikram	Village-Dolu,Raitya road	-
4	Rajendra	Village-Dolu,Raitya road	-
5	Sultan	Village-Dolu,Raitya road	-
6	Vinod	Village-Dolu,Raitya road	-

Annexure-V

Crop	Grapes Orchard Activities Calendar											
	January	**February**	**March**	**April**	**May**	**June**	**July**	**August**	**Sep-tember**	**Octo-ber**	**Nov-ember**	**Dec-ember**
1	Levelling+ Ploughing+ Pruning+ Weeding+ Spraying+ Manuring+ Preparation of irrigation channels	+ Apply irrigation Apply spraying (Fertilizer)	Leaf appearance + Spraying +Manu-ring Apply irrigation +	Spraying + Apply irrigation	Apply irrigation + Spraying	Harvesting+ Grapes marketing	Only foliage appearance	Only foliage appearance	Foliage appearance Yellowish colour	Foliage appearance Yellowish colour	Deciduous stage+ Leafless stage	Deciduous stage+ Leafless stage